Dark Visitor

Dark Visitor

The Coming Ice Age

Billy J

iUniverse, Inc.
New York Lincoln Shanghai

Dark Visitor
The Coming Ice Age

All Rights Reserved © 2004 by Billy T

No part of this book may be reproduced or transmitted in any form or by any means, graphic, electronic, or mechanical, including photocopying, recording, taping, or by any information storage retrieval system, without the written permission of the publisher.

iUniverse, Inc.

For information address:
iUniverse, Inc.
2021 Pine Lake Road, Suite 100
Lincoln, NE 68512
www.iuniverse.com

ISBN: 0-595-30364-1

Printed in the United States of America

Contents

Agent's Preface vii

Introduction 1

Part I—Historical Section

Author's Introduction
Chapter 1   George and his Father 5
Chapter 2   Karen and her Mother 15
Chapter 3   Jack and his Studies 26
Chapter 4   Amiel and his Travels 40

Climatic Effects and Information
Chapter 5   Normal Climate (before dark visitor) 44
Chapter 6   Ice-Age Climate (after dark visitor) 54

Part II—Jack's Report

A Clairvoyant Fable
Chapter 7   Methodology of Discovery 75
Chapter 8   Dark Visitor Candidates 85
Chapter 9   Analysis Model / Accuracy 105
Chapter 10   Coordinate System 116
Chapter 11   Discussion of Results 120
Chapter 12   Graphical Results 130

Appendix
Appendix Introduction 137
Section A1   First Model / Comments 139
Section A2   Three-Body Model (in Part) 151
Section A3   Spawning Other Universes 155

Postscript   Hidden Easter Eggs 157

Agent's Preface

I noticed the following classified ad Billy T placed in *O Estado de São Paulo* because it was in English, not Portuguese.

> **Legally Earn a Fortune**
> **Take CD to USA, S3l7**

I was curious and going anyway, so I left a response in mailbox S317. Billy T came to my apartment the next evening. He gave me the CD. It contained the text of this book, except for this preface, which I am adding. We signed an agreement that gave me the author's rights, until he returns to claim them. That document required me to promptly publish his book. Billy T insisted that the book is a historical record with a scientific report, not a science fiction story. I am not able to evaluate this claim, but as agreed, I am publishing the book/report and will let the reader decide. The historical part tells how his friend (a professional astronomer) came to discover the "Dark Visitor" through its gravitational effects. (It can't be seen, hence its name.) **The scientific report claims:**

A "Dark Visitor" of mass 2.2 times that of the sun is approaching our solar system. The Dark Visitor is probably a small black hole, as it is not reflecting sunlight. Even though the Dark Visitor will be more distant than Saturn's orbit in 2008 when it passes the earth, the earth will be thrown into a new orbit, approaching the sun 6% more closely than the present orbit during the Southern Hemisphere's summer. Evaporation from the oceans will be greatly enhanced, and clouds will cover much of the earth. Southern Hemispheric residents will be subjected to torrential rains. **Millions will die in floods** that sweep away whole cities. Only rice will grow, but life will still be possible there.

Residents of the Northern Hemisphere will have a different problem, not initially serious, but one that quickly turns into an **unprecedented disaster**, far worse than the southern floods. During the northern winters, water that is being evaporated from the warm southern oceans and covering much of the earth with clouds will fall as snow in the Northern Hemisphere. Northern Hemispheric winters will be milder in the new orbit because the earth is closer to the sun, but heavy snows will fall almost every day, much like the great snows storms that sometimes come in late winter or early spring when the winter weather is mild.

In the extreme of Northern Hemisphere's summer, solar heating will be reduced by 19% because the sun will be 11% farther away than before the dark visitor came. The great snowfall accumulations of the previous winter will not entirely melt in these much colder summers. Predictions, presented in this report, are that a **new glacier age** will begin to lay siege to the Northern Hemisphere. Within two decades following the passing of the dark visitor, **Washington DC** and latitudes further north will be buried under thick ice and uninhabitable. The growing ice sheet will reflect more sunlight back into space and accelerate its own spread southward. In less than 100 years, only the Gulf States and the southern parts of New Mexico, Arizona and California may escape the grasp of thick continental ice. Climatic predictions are uncertain. Perhaps only the Key West Islands and Hawaii will escape. **All existing ports will be useless** because the oceans will have vacated the continental shelf, as they have in past ice ages.

W. R. Powell, agent, São Paulo, Brazil

© Brazilian National Library Copyright number 290.331 (Livro 525, Folha 491)

Introduction

I am a historian. History is not just dates and facts. I strongly believe that history must be told from a human point of view. The people must be real to the reader. He must feel their pain and experience their joy if the history is to be meaningful to the reader. It is my privilege to tell the most important story of all time. This is because I have been Jack's friend since the days when we shared an off-campus apartment at Harvard.

When his father died, Jack returned to South America before finishing his Ph.D. research. He had to manage the family's cattle ranch. He was an advanced astronomy student and was able to leave with a Masters Degree, but not the Ph.D. he wanted. After a year, Jack sold the cattle ranch and began to build himself a small, but modern, observatory high in the Andes Mountains. For several years after that we almost lost contact with each other, except for an annual exchange of Christmas cards.

I was very surprised when he called to tell me he wanted to visit me again, as soon as possible, this time with Karen. She was pregnant, wanted to go shopping, and to visit her family. She planned to stay in Boston until the child was born. He said he had "an urgent project for me" that he could not discuss over the phone. He had visited me once before, while he was still single, just after selling the cattle ranch. That time he told me that he had nothing to do while the observatory road was being constructed. He claimed that he needed to return to negotiate the purchase and shipment of his main mirror and other equipment for his observatory, but this was not true. He could have done all of that from South America. He now admits to the real reason for his first trip back.

That first return to Boston was successful, reversing the failure he had experienced 18 months earlier. Karen was in the middle of her junior year at Radcliffe when Jack's father died. He had tried to persuade her to drop out, marry him, and go to South America as the wife of a cattle rancher, but failed. She wanted to go, but decided to remain and graduate, as she had promised her mother she would.

Karen was the third member of her prominent Boston family to be "Radcliffe Girls," and that promise was given to her mother in the hospital where she lay, dying of cancer. They married within a month of his return. I was best man, and will be Godfather to their child. I feel I am part of their family again.

Karen's father is very rich. He had also discouraged Karen from going when Jack first asked. He liked Jack as a person, but correctly predicted Jack would not do well as a cattle farmer because, except for Karen, "his interest was in the stars." Also he did not want his only daughter living so far away and was sure that they would soon be poor as Jack was already telling of his grandiose plans to build his own observatory and live high in the mountains. Karen's father knew that "astronomers never get rich." It is one of life's minor ironies, that Jack's discovery of the dark visitor will make him nearly as rich as his father-in-law, but Karen's father will share in the wealth because he formed a syndicate of his wealthy friends to help Jack exploit the dark-visitor discovery. His share of syndicate profits will make him considerably richer, but Jack and Karen's share is larger.

Unfortunately, I must be content with the exclusive privilege of telling how it all happened. That is what you will learn in the remainder of this book / report. Jack wrote the "report" part of this book himself and I have only lightly edited parts for better clarity to the reader. He provided me with all the graphs found in his section. I thank him warmly. I thank Karen also. She assisted me by frequently reviewing my work, while waiting for her son to be born. She has been very open. She only restricted me from telling how her mother became a Rockette and about her still living maternal grandmother's wild youth.

Except for the names of some of the principals, and some minor details I invented in the historical narration to fill in gaps and flesh out the personalities, this is a true story. I am sorry to say that the global catastrophe reported in this book is coming. Don't blame Jack or me.

The scientifically knowledgeable reader may want to begin reading with Jack's report (Chapter 7), or the climatic effects section (Chapters 5 and 6). If he can follow the mathematical proof given in the first few pages of Jack's report he will know that the dark visitor is irrefutable real. If you are a non-scientific reader, begin at the beginning and, if you are religious, conclude with a prayer.—Jack is confident that the astronomy is right, but the climatic consequences are less certain. Prayer may help.

Part I

Chapter 1

▼

George

Karen was an "unexpected blessing." She has an older brother, George, who is 13 years her senior. Their father, Mr. Goldwater, inherited wealth from his father, who continued to misspell the family name "Gildwaser" in conformity with his father's original Ellis Island immigration document. The first Mr. Gildwaser married a six-years-older spinster whose still living father owned a small fleet of merchant ships, but he profited very little by this. He died in the winter of 1884/1885, more than twenty years before she did. Their son, the second Mr. Gildwaser, was 31 when his mother finally died (1905) but he had had control over the fleet, if not actual ownership of it, in the last decade of her life. He correctly foresaw the potential danger of the technology Germany was developing for U-boats. (More details in Chapter 4.) If used in war, they could destroy many ships. He did not want his mother to continue owning the merchant fleet her father had assembled. In the last few years of his mother's life, she would sign any paper he put in front of her. He sold her ships, one or two at a time, rather cheaply to get rid of them quickly, but he did it in a way unique for that era.—Sale with lease back.—She never knew they were sold. More details are given in the next chapter.

He also foresaw the growth in the oil demand the "horseless buggy" would produce. When his original merchant ship leases expired, he leased small oil tankers instead of exercising his option to renew the old leases for five years more. He enlarged the family fortune by importing oil in leased oil tankers during W.W. I. The capital he gained was invested mainly in the stock market after W.W. I ended. His amazing luck or foresight continued to serve him well. He got out of the market early in 1929. Although he controlled a substantial fleet for 55 years, he liked to quip: "Never owned a ship, not even a row boat," but he made more money than most who did. He died in 1952, 78 years old, while working at his desk, with his silver flask open and a cigar in his hand, according to the family legend.

When his father died, Mr. Goldwater was only 19. His mother, who had worked in the office, at least intermittently for many years, helped him take over the family firm, but he soon discovered that he did not have his father's knack for negotiations with shipowners nor the inclination to travel for negotiations when he could afford to spend his time more pleasantly in Boston. Also the nature of the oil-transport industry was changing rapidly. Most small, privately-owned, oil tankers no longer ventured far from their harbors. They were under long-term contracts to major oil companies and used only to off load oil from the modern leviathans which were unable to come into many ports. Like their smaller cousins, the tugboats, these older tankers now were condemned to monotonous routines of servitude to large and cumbersome giants. Tugboats were made for their task and proud of their power. They tooted their horns as they worked, but the little tankers, who once freely roamed the seas, showed their depression with rust stains and flaking paint.

As a consequence of the changing economics of oil transport, Mr. Goldwater mainly bought and sold oil delivery contracts during the first years he managed the family firm, but slowly he began to diversify. He began to trade in other commodities. After a few years, he ceased even trying to deal with the few, usually contumacious, shipowners who still were clinging to their more glorious past and offering their aging fleets of little tankers for short-term lease. The firm's slogan, "Oil Merchants to the World," quietly disappeared from the company stationary.

To put it politely, the young and wealthy Mr. Goldwater was a not shy man. Now that he is 70 years old, he admits that he was quite a playboy in his youth. Financially, however, he was initially very conservative. He used the futures market only to hedge his oil delivery contracts. During the first years he managed the firm, he never speculated in it, but he was successful as commodities trader. As he acquired experience and noted the fortunes being be quickly made by some, he began to take diverse positions in a wide spectrum of commodities. Oil, the commodity he knew best, made up an ever-decreasing portion of his portfolio. Despite this, everything continued to go very well for many years because his portfolio was usually diversified, but one spring, with one third of his portfolio short in pork bellies, he was trapped by a late freeze, which killed Iowa's corn sprouts as it swept through the midwest. Thoroughly chastised for his unfounded pork-belly enthusiasm, he took his enormous losses and returned to his original cautious approach. He began to focus his efforts upon agricultural grains, eventually dealing exclusively in soybeans. Again the firm dealt in only one commodity.

"Protein for the World" appeared in big gold letters on the wall behind the lobby receptionist, but no slogan was added to the stationary.

His "pork-belly lesson" as he puts it today, changed him completely, not just his financial behavior. The financial losses of the pork-belly fiasco also transformed his wife, whom he had hastily married a few months earlier, but that story properly belongs in the next chapter. In an effort to regain the fortune he had lost, he became what is now called a workaholic, studying everything that could effect the prices of soybeans with great care.

Even when oil dominated the firm's activities, the government's periodic forecasts for the coming winter drove the routine market. There were of course more violent price swings, primarily associated with strikes, wars or unexpectedly prolonged winters, but now that he was specializing in soybeans, creditable forecasts for near-term changes in climatic trends, especially relating to late freezes, droughts or excessive rains, were his primary concern. The firm hired a full-time weatherman, not to forecast weather, but to assist him evaluate available long-range weather forecasts.

Mr. Goldwater was not only an unusually cautious soybean trader; he was also naturally inclined towards long-term planing. On his tenth birthday, George was expecting to get a bicycle from his father. Instead he got an expensive weather station. George got the bicycle also, but that was his mother's present. This weather station was not just a recording barometer, wind / rain gauge and a high-low thermometer. It was a weather radio integrated with a laptop computer that drove a single-pen horizontal plotter. This system could generate the current weather maps as well as detailed forecasts for selected areas.

To put it mildly, George was not thrilled by his fathers gift, but on rainy summer days and during winter evenings after his homework was finished, he began to use it. George soon discovered that he could ignore "bed time" for more than an hour if he was working with his father's gift. Slowly he became interested in storms, especially the hurricanes that moved up the North Atlantic coastline and threatened Boston. By the time he went to college, he was addicted to following unusual weather phenomena occurring anywhere on earth, just as his father had planned. Mr. Goldwater wanted George to become an expert in near-term climatic trends. He expected George to join the family business and head up the firm's section that evaluated the available forecasts when he graduated. His efforts were only partially successful.

George did become a climate expert, as Mr. Goldwater had hoped, but when he got his Ph.D. he did not join the family firm. Instead, after graduation, he went to work for NOAA (National Oceanic and Atmospheric Administration) as a climatic modeler / researcher. Most of his work was concerned with modeling the effects of small changes in the ocean circulation patterns upon global climate trends and conditions. (El Nino, Gulf stream flow variations, etc.) However his expertise in modeling, added variety to his work. He was often absorbed into taskforce teams, working on special projects. With one of these teams he worked exclusively on the "ozone-hole" problem for several months. Another team was concerned with "global heating." It lasted for more than two years.

George is modest, but I surmise that he lead the subgroup effort of the global heating team related to the effect of oceanic absorption of carbon dioxide, as modified by wind-driven changes in the "mixing layer." I don't really understand what these words mean, but George and I get together for a few beers each time he returns to Boston and he likes to talk about his work. I am very interested in how the government approaches politically sensitive problems so I encourage him, but most of the technical details go over my head, especially after the second beer.

George's father routinely consulted with his son before establishing long-term positions in the soybean market. He did this more to try to interest George in the firm and its activities in the futures markets, than to gain any information. The firm now had two experts working full time to evaluate the various climatic forecasts available. They were better informed about current conditions and predictions than George. He mainly refined his computer models. Sometimes George traveled to South America to collect more detailed oceanographic data for NOAA about recent El Nino surges. At least twice he went to sea with coastal fishing boats and periodically collected hundreds of water samples during each of these trips for salinity analysis etc. back in Miami.

It was while George was returning to Miami from one of these trips that he met Jack, who happened to be seated next to George on the airplane. Jack was returning to Harvard after his summer vacation. There were no direct flights to Boston. Jack always changed planes in Miami. Jack was just beginning his Ph.D. studies, but he already knew that he wanted to do his Ph.D. research on storms in planetary atmospheres. At that time he was planning to try to model the giant red spot of Jupiter as a long-standing hurricane. He noticed that George's laptop had a NOAA property tag on it. When George closed it, Jack initiated a conversation that lasted through the remained of that long flight. When they parted, George gave Jack his card and wrote his father's phone number on the back of it. He invited Jack to discuss their common interest in modeling fluid circulation effects upon weather phenomena during the coming Christmas vacation when George would also be in Boston.

I had met Jack at Boston's LAFC (Latino / American Friendship Club) the previous spring. After a few beers, we decided to rent an apartment together for the next school year. I found a nice one, close to campus, the next day. I had to be out of the university dorms in three weeks and was glad when Jack liked it also. I remained in the apartment all that summer while Jack went home to help with the family's cattle ranch. His father was not in good health and wanted Jack to stay or at least transfer to a closer university. I never met Jack's father, but I gather that their relationship was much like the relationship between George and his father had been while George was still in school. Both fathers assumed that their sons would affiliate the family business after graduation and ultimately extend its life for another generation. Neither son actively refuted this assumption, but neither was looking forward to it either. To avoid telling his father a little longer that he would not be joining the firm, George traveled throughout the South and West of the US after graduation and went to Europe in the spring of 1994. He joined NOAA that fall.

Apartments close to campus were not cheap and had to be rented for 12 months, whether or not they were used during the summer. I offered to pay a little more because I would be using the apartment all year, but Jack would not hear of it. He said he thought he should pay me to remain in the apartment and keep an eye on his stereo etc. We both knew Jack was much better off than I was financially. I jokingly replied that I would do it for free if I could use it.—It was a privilege to listen to his records and CDs on a really good stereo system.

When Christmas vacation came, we had been living together for several months and had become good friends, despite our academic differences (Snow's "two cultures"). That Christmas Jack did not return to the cattle ranch during the vacation as he had done every year before while he was still an undergraduate. He had just been there in November for his mother's funeral. She had died suddenly with no preceding illness. They buried her in the family graveyard, near the tree her mother had planted as a child. No autopsy was done. Jack's father did not want anyone "cutting on her" when there was no need for it. Everyone just assumed it was heart failure. Before he returned to Harvard, Jack continued the family tradition by planting a tree in the family graveyard—a small hill, fenced-off from the vast valley of grass and grazing cattle. Although he has sold the ranch, Jack still owns that hill. It will stay in the family "so long as there is a tree upon it."

Instead of going home again in December, Jack called to see if George still remembered him. George did. He invited Jack to spend the week between Christmas and New Years at the family's country estate in the Green Mountains. Jack was reluctant to leave me all alone for a whole week during the holiday season; (I had no place to go. My family was killed in Nicaragua's civil war.) so he asked if his roommate could come too. George said "the more the merrier" and that the house was rustic but large. The estate also had several small cottages used 60 years earlier by lumberjacks, but now converted into guesthouses, so it was "nooo problem."

When we arrived at the estate on the 26th, George gestured to several of the cottages and said: "See—nooo problem." The way he said it made me think he was making fun of my accent, but I soon learned that he used that same expression with everyone. Despite the extreme difference in our backgrounds and financial positions, George and I have become good friends, drinking buddies one might say. I have been back to the estate once. George took me turkey hunting just prior to Thanksgiving, but we did not see any turkeys. I was almost as glad as his father was when he returned to live in Boston, but I am getting ahead of my story when I tell this.

That is how I got to know the Goldwaters and how Jack met Karen. None of us thought at the time that George's expertise in climatic modeling would be important to Jack. That came years later, after Jack had discovered the dark visitor. George has a share of the investment syndicate, granted officially as compensation for his work in making the climatic predictions about what will happen after the passage of the dark visitor. Chapters 5 and 6 give his main results and explain the basic concepts used in his detailed computer analysis.

I was already working on my thesis, which concerned the changing rights of women within South American legal systems when I met the Goldwaters and several other houseguests they had invited for that Christmas—New Years week. George's sister, Karen, was a freshman at Radcliffe University when we met her that Christmas vacation. Radcliffe was developing a "woman's studies program", as were many other universities at that time. I was already going to their library at least once each month, because in my narrow field it was better than Harvard's (at least, more modern as library funds were being spent in support of the new program area). After meeting her, I often had lunch with Karen during these visits, but it was Jack she invited to the spring festival weekend. Like George, I am becoming a confirmed bachelor. I hope the reader will excuse these personal digressions, but I am also part of the story as well as the narrator.

Mr. Goldwater continued to hope that George would return to manage climate-forecasting activities for the family firm. To make this more a more attractive proposition to George, the firm had purchased several years of hourly global weather records stored on old 9 track tapes and the two floor-standing tape drives plus an old IBM 7094 computer that came with them. Mr. Goldwater got them during a GAO auction of surplus government property for less than the cost of transporting them to Boston. Most of the data had been transferred to CDs, but the complete set of corresponding CDs was much more expensive. The three machines and thirty boxes of tapes nearly filled a moving van. The tapes alone now occupy two small rooms in the firm's basement.

One of the two full-time employees who is responsible for evaluating climatic forecasts has integrated this very extensive database with the firm's more modern computer system, but no one is using the information. The volume of weather information archived but potentially available to the firm's computer system now greatly exceeds that about soybean trades and other market related information. Let me describe Jack and Karen's wedding to tell you how I know this.

Jack and Karen's wedding was held Sunday afternoon in a small "church of convenience" only one block away from her father's office building. Neither Jack nor the Goldwaters were religious, but a certain amount of pomp and ceremony was required. The event was well covered in the social pages of the Globe. Before the wedding ceremony, Jack, George and I were sequestered in the firm's basement where the computer center is located, while Karen and her bride's maids dressed upstairs. My job was to try to keep Jack relaxed and George's unstated one was, I think, to keep an eye on us in this expensive, usually off limits, facility. Soon Mr. Goldwater came down. He said he wanted to give Jack and me a detailed tour of the facility, while we waited. It would help everybody relax, he said. I think the tour was really intended for George. George had been avoiding going to his father's office for years.

It is a very impressive facility—Except for two couches and a coffee table near the entrance, the main room looks like a small version of NASA's flight control center. CNN and Bloomberg data were on separate large-screen displays high on the wall to the right of the entrance. Two other large-screen displays on that wall were blank as it was Sunday, but during the week, they would have streaming data from various commodity and stock exchanges. The main computer was actually four modules that looked identical, but one was "master" and the others were "slaves." They were small table-sized units on an elevated section of the floor, like Gods on a pedestal, Jack said. Also on the elevated floor section (one step up on the left side of the room) was a glass-walled subdivision which Mr. Goldwater called his command center. In it was a central table with about 10 phones and half a dozen computer keyboards with flat screen displays in front of them. The desktop type computers were under the table with some other metal boxes which I think made requested outputs from the main computer available to these command and order terminals.

The two floor-standing reel-to-reel tape machines and the old 7094 computer that serviced them were on the far wall, at regular floor level. They looked out of place amid the modern machines and glass architecture. Next to them was a door. It open into the hall that led to the tape storage rooms behind this old GAO surplus equipment.

In the center of the room, but near the entrance, was a long low table made of chrome and glass with leather upholstered couches on each side. After the tour, we sat there. Most of the talk was too technical for me to follow all that was said. It was something about the older machine being tied into the network for "parallel processing" by the four modern machines. These later generation computers were much faster and much more expensive than his GAO bargain. They would do any analysis and modeling George might be interested in doing. The old GAO equipment would just feed them data as needed. I think the point was that, if they were processing the old weather information, the large volume of input data required from the tape machines would limit the system's processing speed. It could be a "background task" that would not interfere with market analysis or any of the firm's normal activities etc. Mr. Goldwater was obviously proud and pleased with the computing power he had assembled at relatively low cost by networking four modest modern machines. He seemed to really understand it too, and that impressed me.

One thing in the room impressed me more. It silently confirmed a lot of things Karen had told me about her father during our lunches at Radcliffe. It was a small, glass-topped, mahogany table—a display case, pushed up against the back wall, near the right corner of the main room. Karen had told me about George's tenth birthday present once when she was complaining about how her father was "guiding" her and had manipulated George. That glass and mahogany case contained an archaic horizontal plotter and an old laptop commuter connected to a small radio. The case's brass nameplate simply said: "Starting Point." I don't think George knew it was there. I had noticed that it was the only wooden thing in the room and was curious to see what it was.

When I rejoined the others, I understood that Mr. Goldwater was again asking, almost begging, George to join the firm. He said it was a shame no one was making use of the database he had "foolishly assembled." If George did not want to use it, perhaps he should at least try to sell the tape drives and the tapes storing the data. George appeared to take this threat seriously and discouraged him. Saying the data was much more valuable than the tapes. They included a great quantity of ocean and atmospheric data plus most of the US government's prior climatic forecasts. Now that they were integrated with the firm's financial data, the commodity market's trade-by-trade reactions to government releases could be studied. George said that this integration of market and weather data was a valuable and unique collection and should be preserved.

Naively, he asked his father why he had collected all this old weather information. "Was he planning to join the competition between NOAA and the US weather bureau?" No, Mr. Goldwater replied, he did not want to use the database to make better weather or climatic forecasts or even to evaluate existing forecasts. Instead he wanted the firm to build a model that would take the latest data available and predict the content of the government's next long-range weather and crop-yield forecasts. I can still remember him saying: "I'd rather know what the government is going to forecast than to know what the weather is going to be."

George said he was currently revising a model that used parallel processing for millions of synthetic weather records because NOAA had a similar data-input problem when using real weather records. When working with actual weather records, instead of internally assumed conditions, the large volume of input data required often limited NOAA's machines. He seemed genuinely impressed with the facility his father had assembled.

After the wedding we returned to the same building, most of us walking, because it was now cool and the air was fresh from the earlier rain. The reception was a very plush catered affair, fully occupying the spacious lobby. Everything had been removed, but "Protein for the World" was still on the wall above the table where the cater had set up the bar. It struck me as funny at the time, perhaps because I had already had a couple of drinks before I noticed that the bar was exactly centered under it.

George and I both never strayed far from the bar. After a while, Mr. Goldwater came over to us with Karen and Jack in tow. Karen said there was a shortage of men and we should mix a little more. Mr. Goldwater ignored this and immediately asked George if he had thought any more about the database he had collected. I don't know if it was the drinks we had had, or the fact that George would be returning to Miami the next day and Karen was soon to depart for South America, leaving Mr. Goldwater all alone, or something else, but we were all dumfounded when George said that it (the database) had some "interesting potentials" and then announced that he was now seriously considering joining the firm as his father had long wished.

Mr. Goldwater was already in a good mood. Everything about his daughter's wedding had gone smoothly and he had had a drink or two, which was unusual after his pork-belly transformation. When he heard these words from George, Mr. Goldwater initially said nothing, but the Mona Lisa smile, which spread across his usually taciturn face, spoke volumes. Soon he said that he was still regretting the fact that Karen was going so far away, but it would be "some compensation" if George did return to Boston. In any case, about four months later, George resigned from NOAA and joined the firm. While George was driving back to Boston, Jack was carefully hauling his mirror up the mountain. Discovery of the dark visitor was still more than a year away.

Once his father's 23-year campaign to get George into the firm was over, he started on a new one. He wants George married so the firm will have an heir for the next generation. He knew Karen would not leave Jack and Jack would not leave the observatory he was building. New England's light pollution level, frequent clouds and mountain fogs made that unthinkable, even if Mr. Goldwater did build him a better observatory in the Green Mountains of Western

Massachusetts as he once had offered to do. So that left George as his only hope to perpetuate the firm under family control.

George probably is still one of the ten "most eligible bachelors" in the Boston area. When George's mother was alive, George escorted several of Boston's high society girls to debutante balls in the spring, but nothing serious ever developed. He seems content to remain a bachelor, and concentrate on his work. George is president of the dark-visitor investment syndicate, and chairman of the investment committee. He makes many decisions alone with the aid of his computer programs and acts upon them immediately, within the established limits.

His father was initially ecstatic with George's progress in the firm, but now he is worried that George is becoming too absorbed by computers and the commodity market—so busy that he has no time for a social life. George and I still occasionally get together for a few beers, but that seems to be the extent of his relaxing time. He is aware of his father's concerns. At our last meeting he told me that his father had hired and attractive blond assistant to work with him. He said that it was "OK", as she is very qualified and helpful, (She has a Master Degree in computer sciences.) but he wishes his father would stop trying to shape his life.

Chapter 2

Karen

Karen's great grandfather fled Russia before he was a teenager with two rubles in his pocket and his father's blessing. His older brother, Amiel, who was charged with the task of protecting him, accompanied him. Their father sent them away with the admonishment: "Seek your own Promised Land.—This is not it." There had been one pogrom too many in their village. Where they went and what they did is not known, but two years later they ended up in Copenhagen. The older brother sold his charge to an elderly Chinese man to work as an apprentice in his apothecary which was located near the harbor, not far from where the little mermaid sits and waits today. He never saw his brother again.

The clientele of the apothecary were a cosmopolitan mix of seamen, resident stevedores, their wives, and members of the local Chinese community. The shop owner practiced medicine from his apothecary as well as selling teas, herbs and other medications. Most clients came directly to the shop with their medical complaints. No prescriptions were required and many languages were spoken. The owner did not know Russian or Polish and welcomed the addition of his apprentice who did as some of the sailors who came to his shop were East Europeans.

Karen's great grandfather lived in a room rented from an elderly Jewish lady who constantly complained of gout. He told his master of her complaint and was given a tea to take to her. Soon she was better. Convinced of the effectiveness of the old man's medications, she and some of her fiends became regular customers of the apothecary. This inroad into the Jewish community also made the old man pleased with his young apprentice. Karen's great grandfather continued to work in the shop after his four years of indentured servitude expired. The old man had become a substitute father for him and he a son to the old man. Four years later, in August of 1864, the old man died. No one was surprised to learn that to Karen's great grandfather was named in his will as the owner of the shop and the narrow three-story building which housed it.

Six weeks later, he married the niece of his former landlady and the young couple moved into the second floor. In that first winter, they seldom went into the third floor, because it was full of junk the old Chinese pharmacist had collected over the years and the roof leaked. It was necessary to go there relatively often during the following spring to empty the strategically placed tubs, which protected the second floor. The accumulated dirt and clutter began to annoy his bride. Even though she was six months pregnant she began cleaning and clearing out the third floor. That is when they discovered a small chest full of jade carvings and approximately forty small gold coins, mainly from the orient. Not sure of their title to this treasure, they told no one, but began to display and sell their inexhaustible store of jade in the shop, a few pieces at a time. The gold coins they kept.

Three months later, his bride of less than one year went into protracted labor and died. The tragedy was compounded further by the fact that his son, when finally delivered, was still born. The pharmacy business was failing. Many previous customers came mainly for the old man's medical advice. Local Chinese customers complained the he was misrepresenting weeds as medical herbs or selling mixtures that contained more weeds than herbs. This was probably true, as the sailors who supplied them to him must have realized that he only knew the price he should pay and could not evaluate the quality of the "herbs" he was buying from them. Even with the profits from an occasional jade sale, he was losing money.

He soon decided that the Denmark was not the Promised Land either. In September of 1865, he sold the remaining jade to a Chinese merchant at a great discount, closed the shop, sold the building at a fair price, and sadly set sail for New York. His depressed state of mind must have been tempered by the bank draft, payable in dollars, he carried and almost forty gold coins sewn in his coat and pockets. He celebrated his 22nd birthday on the boat and was three days sail from New York during Yom Kippur. After the requisite examination of himself, his spirits must have brightened. He was barely 22, strong and healthy. He probably thought to himself: "I am much better off than when I left Russia with only two rubbles in my pocket."

He had considered returning to Russia to try to find his real father but the tsar was still on his throne and pogroms were now district-wide government-supported affairs, no longer isolated incidents in scattered villages. He had written to his father a year earlier, when he inherited the pharmacy and married, (care of the rabbi of his village), but the letter returned a few months later with a stamp and note telling that no village with that name existed in that district, so he decided his dead mother's brother in New York City was a better bet to learn what had happened to his father. The US civil war was over and people were saying that America was the Promised Land.

The above facts all come from a small diary, probably one of his wedding presents, as entries relating events prior to the wedding are all made with the same ink. I have a friend in the language department who was able to translate most of the diary for me, not an easy task. (Karen' great grandfather had little or no formal education. He wrote in a strange hodgepodge of Russian, Danish and very corrupted Hebrew, usually everything right to left and phonetically spelled.) Unfortunately, he did not describe what he did in New York or where he lived. The last part of the little book remains blank. None the less that little book is a historian's treasure. Working class diaries from the nineteenth century are very rare. Most workers were illiterate and the few who could write had no money for diaries. I have given the original, along with the translation, back to Karen. The translation and photocopy will serve my needs. The last entry, probably written on the Ellis Island Ferry to Manhattan, is short but poignant. In translation it states: "Praise God—A New Life in New Country and <u>a New Name</u>!" (Underlining in the original)

What happened during the next seven years is unknown. Perhaps some information exists in NYC, but there is not sufficient time to investigate, as Jack wants me to finish telling this story quickly. When I learned that Jack knew the pending fate of the earth and was not telling anyone except Karen and George and a little later, when George's preliminary analysis was done, Mr. Goldwater, I was angry with him, initially. I now understand that secrecy was necessary for the syndicate to establish its positions in food and oil commodities at prices that will greatly elevate once the public is creditable informed. Jack thought that if only his text were presented, especially if presented directly by him, that few would read it and it would be dismissed as science fiction. If this happened, the syndicate would have to sustain the interest cost of their investment for several years—until the dark visitor actually begins to significantly disturb other planets in their orbits.

Most of the silent partners of the syndicated will learn the true basis of their investment from this book. Their participation was based on Mr. Goldwater's guarantee that at least their capital was safe. I doubt that many would have participated if they knew the current facts without this guarantee. I don't think Jack is wrong about the dark visitor, but he could be. Because the dark visitor is still very distant from Pluto, most of the small disturbance it has produced while approaching the orbit plane of Pluto has been attributed to other causes. The cumulative effect of the dark visitor to date is less than what was observed once when an asteroid scattered out of the Kuiper belt (or Oort cloud) and passed relatively close to Pluto. Because of these perturbations, some people have suggested a tenth planet, planet X, exists, but few professional astronomers support this view. Jack's collection of detailed observations made at Harvard's Center for Astrophysics, CfA, by himself and others during the period of meteor bypass acceleration and reversal can not be repeated at present. Only the residual displacement remains which is a combination of both the brief meteor and longer-lasting dark-visitor perturbations. This residual combined effect can not be used to compute the trajectory of the dark visitor. The short-term "impact" effects must first be removed to see the currently still small, but cumulative effects of the dark visitor. How to do that and the data collection necessary that he has assembled is knowledge Jack is keeping secret until he abandons hope of actually seeing it.

Jack tells me that some theories of small black holes predict that even black holes emit some light from fundamental particle events occurring in the intense gravitational fields in regions just outside of their "no escape" boundary. Perhaps his knowing reasonably accurately where to look will let him see this light, if it exists. If it is a black hole, it will never be seen in reflected sunlight the way the moon and other planets are seen. Even if a black-hole dark visitor had a small, super-strong, highly-polished, spherical-mirror, surrounding its central point, the reflected light would keep returning to the mirror surface until it was absorbed instead of escaping. There are other possibilities, extremely dense objects, that do reflect sunlight. These are all discussed in Chapter 8.

Except for Neptune, all the other planets are much more tightly bound to the sun than Pluto is. That is, most planets are constantly subjected to solar forces larger than the force the dark visitor can exert upon them until it approaches their orbital planes. The dark visitor will pass behind Pluto and closer to the sun when it crossed Pluto's orbit plane. It will not eject even weakly bound Pluto from the solar system. Instead it will slow Pluto down and cause it to fall towards the sun. Jack can not precisely predict how close Pluto will come to earth, but assures me that Pluto will be always be very far from earth (at least 15 times more distant than the sun). It will never hit the earth. If the dark visitor were to pass in front of Pluto and a little more distant from the sun than Pluto, it could eject Pluto from the solar system, but it will not.

It will only slightly effect Neptune and Uranus, because when it passes through their orbit planes, they will be on the other side of the sun. It will pass much closer to Saturn than any other planet. The orbit of Saturn will be strongly effected and it will also fall towards the sun as a result of the dark visitor. Saturn will be a new spectacle in the sky in a very elliptical orbit. At times during its new orbit it will be by far the brightest "star" in the sky. Jack will not let me be more specific than that and is providing no graphical information about Saturn. If

Saturn is not ruptured, Jack says that it too will never hit the earth. (The interaction with Saturn is so intense that even its orbit plane will be tilted several degrees more away from the ecliptic.)

There is a remote possibility the dark visitor will exterminate all life on earth but it can not do so directly. In addition to drastically disturbing the orbits of Pluto and Saturn, it will disturb many asteroids in the Oort cloud. (Many astronomers consider that Pluto itself is one of these asteroids, not a true planet.) Many of these asteroids will be violently disturbed and ejected from the solar system. Many others will be thrown into highly elliptical orbits, and pass near the sun. If a large one should hit the earth (an extremely unlikely, but possible event) then the dark visitor will have killed us all and not "just" have made most of the Northern Hemisphere uninhabitable several decades from now.

My assigned role in this project, for which I have received exclusive publishing rights and full cooperation from Karen and Jack, is to quickly augment his report with their story. Both to make it more interesting to the non-scientific reader and to provide historical facts that can be checked. (In his section, Jack has provided the scientifically knowledgeable reader with quantitative arguments in support of the dark visitor using well-known facts.) I am happy to do this as this book should sell well and permit me to leave Boston before it freezes over in about a dozen years. In addition to their financial rewards, the silent partner participants (with emphases on "silent") in Mr. Goldwater's syndicate will receive handsomely bound editions of this book. Then they will understand why Mr. Goldwater could afford to pledge his personal fortune to assure them a risk free investment, if they all keep their mouths shut about the existence of the syndicate while the investment is quietly made. Now to resume Karen's history.

Boston rabbinical, commercial and newspaper records all show that Karen's great grandfather married Zelda Swartz who was five or six years his senior but the daughter of a wealthy shipowner. Contemporaneously he became the co-owner of one of his father-in-law's ships. A wedding present (or dowry bribe?). Zelda was at least 34, probably a few years older at the time of the marriage. Karen's grandfather was born only 13 months later so the wedding was not entirely for commercial reasons. What happed to the gold coins is not known. Although co-owner of a ship, he could not sell it. As far as is known he had no job. Other facts seem to indicate he had no income from any source. His wife and her father were not known for their generosity, so he may have slowly converted the coins into beer and tobacco for his pipe. He avoided the house, spending most of his evenings in harbor bars.

How he met his wife is pure speculation, but the following is known. Zelda's father had a brother who was a rabbi in NYC. The Boston Jewish community was much smaller and most of the single men did not return after the civil war.

Probably, if Zelda had not found a husband there, despite her father's wealth, by the time she was 30, her father may have sent her to NYC for prolonged visits with her uncle, who knew and could influence many eligible orthodox bachelors. Karen's great grandfather died of "consumption" during the winter of 1884/1885 with no more than 43 years of age. He may not have been a happy man. Perhaps his most pleasant years were his period of indentured servitude and the years that followed as the old Chinaman's unofficially adopted son. Surely, he never found the Promised Land.

Karen's grandfather, the second Mr. Gildwaser, was not only a good negotiator of contracts; he was also a financial genius. When he took over control the family fleet for his mother, he sold all of the ships and leased them back for four years with option to renew for five years more. While this is common now in many industries, it was a very innovative move at the time. This not only gave him a great cash windfall, which he multiplied many fold in the stock market of the early twenties, getting out before the 1929 crash, but it also gave him the flexibility to least different types of ships when the original "lease back" contracts expired.

Before the end of 1910, all of his lease contracts were for oil tankers. The contracts contained a clause that required the owner to insure (or by other means satisfactory to him, to guarantee) the replacement value the cargo (oil) if the ship sunk. The owner supplied the captain and crew, but the contracts had a minimum average speed specified. Three of these old contracts still exist and I have read them.

W. W. I began in July 1914. Many shipowners suffered when U-boats began to sink ships, especially owners of the slow and very vulnerable oil tankers. Insurance rates went though the ceiling so even owners lucky enough to not lose any ships were losing money if their ship was under lease to Mr. Gildwaser. Oil was relative cheap when it was being loaded onto the ship, but very dear if it reached the East Coast of the United States safely during W.W. I, where the "replacement value" was defined under terms of his contracts. Sometimes Mr. Gildwaser made more money if a U-boat sunk his leased ship than if he sold the oil, even thought oil demand exceeded the supply. He never had cause to regret that he did not own ships. He was proud of the fact. He not only correctly foresaw the crash of 1929, but also the post-war surge in oil demand as Americans fell in love with the automobile. He is the main source of the family's wealth.

He was Jewish, but unlike his father, he observed only the highest of holy days. He had little interest in anything but the growth of his financial empire. Perhaps this was true because there was no place for him in Boston's society, at least before 1929. After the market crash, he may have been the richest man in Boston. It was increasing hard for Boston's high society to ignore him. ("Ostracize" might be a more accurate way to put it.) The world was changing rapidly. Locally the Wobbles were marching again. Europe was in turmoil. Fascists ruled Italy. Republicans were threatening civil war in Spain. The communists were organizing. Faced with a common threat, the common man, and diminished in numbers, Boston's financial elite were apprehensive. "Law and order were at risk." (Their privileged position and laws favoring the wealthy were changing might be a more accurate description.) Pinkerton men and other private police forces had to be paid for if "social order was to prevail." Boston's society might have a place for their most financially successful and politically conservative citizen at last.

Mr. Gildwaser was 55 years old when the market crashed. Although largely undamaged financially, he was not immune to the uncertainties of the times. Amid the social upheaval he had his own private catharses. When financial empires were crumbling he sought something solid and enduring. He decided there were more important things to life than his work. He realized late in life that he wanted a son. He began to go to synagogue more often, but he did not find what he was looking for there. Instead, he found what he was looking for in the lobby of his own firm. He had recent hired Kathleen to replace his aging receptionist who had returned to Providence for reasons now unknown.

Kathleen came from a socially prominent, but now financially ruined family. She was a graduate of Radcliffe University, but had never worked before. Her very presence in the lobby added class to it. She was attractive, warm and friendly, but she was also Irish, Irish Catholic. Initially they had little contact but business was slow in the post depression years even for an oil importer. The activities that had previously filled his day, often until six or seven in the evening, no longer seem so important. Mr. Gildwaser found that he "needed some exercise", as he put it. He would walk down four flights of stairs to the lobby to see if the mail had arrived or to get the Wall Street Journal. A little short of breath, he would linger and talk to Kathleen before taking the elevator back to his office. Somehow it was hard to concentrate on his work. His trips to the lobby became more frequent and lasted longer. There was little for Kathleen to do. The phone seldom rang and visitors rarely came to the office. Soon they were no longer employer and employee but two intelligent friends telling each other stories to pass the time of day.

Boston society was shocked when their engagement was announced. The bishop refused to marry them. Mr. Gildwaser would not convert nor promise to raise any children as Roman Catholics but he ceased celebrating the holy days. They wed in a civil ceremony late in 1931. Kathleen was only 34. Despite the prognostications of many, the marriage was happy and fruitful. Mr. Goldwater was born in the spring of 1933. Shunned by Boston's society, they delighted in their son. Mr. Gildwaser would often close the office early or return home for a two-hour lunch. Business was slow and it could wait.

No one seems to know why Mr. Gildwaser changed his son's name to Goldwater. Perhaps it was at Kathleen's request to make it less foreign sounding. It first appears on the baptismal certificate and that was surely Kathleen's doing. Perhaps the priest suggested it to reverse the error of the Ellis Island clerk.

When Mr. Goldwater was old enough to go school, Kathleen took him and then often joined her husband in the office organizing his files or other minor but trusted tasks. An elderly lady now occupied her former chair in the lobby. W.W. II began in the fall of 1939. Although the fighting was confined to Europe, oil was contested globally. His work took on a new importance and was challenging again. When it ended, he was 71, but with a young son and wife who was only 48, he did not sense his age. He continued to work long hours in the office for seven years more until one afternoon Kathleen found him slumped over his desk.

Kathleen ran the office alone for half a year so that her son could finish his second year of college. Mr. Goldwater was spoiled as a child and did not want to go to college, but his father had insisted. He preferred the easy life his financial circumstances would permit. Somewhat reluctantly, he assumed the presidency of the firm six months after his father had died. During the day, with his mother's help and under her watchful eye, he quickly master the necessities of the business world, but even working together they were not able to continue the firm's tradition as "Oil merchants to the World." There simply was no longer a place for small tankers and small oil import firms. Even the "seven sisters" (major oil companies) were merging. More of the details of this period have already been reported in the first chapter.

Young Mr. Goldwater liked excitement, but not stress. Speculating in the commodities market is very stressful even for experienced traders, but commodity trading was not stressful for him. He was just hedging his oil delivery contracts. If he lost money there he would make the loss back on the oil contract. He was wealthy already. "Why should he subject himself to stress?" was his attitude. All his life his wishes had been indulged. Even the two years he was at BU were just an extended party. His mother was again welcomed back into the high society that had shunned her after her marriage, especially at charity affairs, when she brought her checkbook, but her son was more inclined to Boston's nightclubs and the Old Howard's shows.

He was a rich and eligible bachelor. The doors of Boston's high society, which were closed to his father, were wide open to him, but he told his mother that he was not interested in entering that "circus." When she complained that he was "running with the wrong crowd" going to the Old Howard (a burlesque theater) where "no gentleman should be seen", etc., he told her that he needed to relax from the high-pressure environment of the commodities trading. His mother had no experience with commodities and never became involved in this new aspect of the business. She believed him or at least pretended to. She had always indulged her son and it was too late to change now. They seldom openly quarreled and soon reached an unspoken truce. He could do as he pleased in the evening, but had to be in the office before the Chicago commodities trading began. This gave him time to sleep late, recover from any hangover, etc. The firm was steadily making money or at least seldom losing any so when she turned 58 in 1955, she ceased working in the office. Just before she quit, she replaced the aging receptionist with an attractive young lady with a "Proper Bostonian" background. It had worked for her—perhaps it would work again.

Considering her own socially rebellious marriage, his mother was in a weak position to protest when at age 31, Mr. Goldwater announced he was marring a dancer he had first seen on the stage of the Old Howard. Initially concerned that her prospective daughter-in-law was a simple gold digger, she hired a detective with instructions to "buy her off, whatever it costs." The detective soon reported that it was not going to be that easy. Like the receptionist she had hired, the prospective daughter-in-law came from a prominent Boston family. Her mother was also a graduate of Radcliffe and was even a leader in the effort to close down the Old Howard. Her rebellious daughter had been a student of ballet for ten years, but a late growth spurt in her mid teens had made a carrier on a more respectable stage improbable, if not impossible. Her ambition to dance in the New York City was transformed by this physiological development, but not abandoned.—If not the stage of the NYC Ballet, then Rockefeller's Radio City Music Hall's stage would have to do. She thought her long legs were made for those high kicks.

In her mind, the Old Howard's stage was just a necessary way station before the "Big Apple." When she was dancing on the Old Howard's stage, during its last years[1], she thought surely someone would recognize her talent and classical training, even though they were satirically displayed in her routine. Someone did. It was the young Mr. Goldwater, then in his late twenties and still very much in his playboy era.—His "pork-belly lesson" came a few years later, just after they were married. I am not permitted to tell how, even though four months pregnant with George, she finally realized her NYC stage ambitions as an unauthorized replacement in the Radio City Music Hall's Christmas show. Although it was

---

[1] She was certainly not their normal stripper. She was probably hired to lend an element of class to the stage, in a failing effort to resist the prudes of Boston, who finally succeeded in closing this world-famous landmark where Gypsy Rose Lee and many others got their start.

only for one night, she was a Rockette! After that night, she never appeared on the stage again.

She reconciled with her mother shortly after George was born (May 1967). Both of Karen's grandmothers discovered they had much more in common that their status as "Radcliffe Girls" and previously ostracized members of Boston's high society. They became close friends. I am revealing nothing not already known when I tell that one hired a detective to investigate the other's daughter and "buy her off." Unfortunately, Karen's maternal grandmother will not permit me to tell her history. She is still alive and now is a very "Proper Bostonian," avoiding any notoriety, but in her youth, she was a wild one. Karen wants me to honor her grandmother's wishes, and I reluctantly do so.

After the financial set back Mr. Goldwater always refers to as his "pork-belly lesson," his wife, Karen's mother, also became more conservative, seldom acknowledging she had danced on the Old Howard's stage, but always quite willing to claim that she was a Rockette in her youth. She died in October of 1999, so Karen has permitted me to tell her story. Karen herself has a rebellious streak, perhaps a genetic gift from her grandmothers. When she was little, she pulled the ribbons from her hair that her mother had carefully placed there. She was something of a "tom boy" in her teens. Karen's mother wanted her daughter to embrace all the things she herself had rejected, especially going to Radcliffe, if possible. Like her father, Karen wanted nothing from the "circus" that was Boston's high society. She refused to "come out" and rarely attended the few debutante balls that still were held in the spring. Her passions were horses and sailing, not gowns and garlands. Dancing, which had so enthralled her mother, was of no interest to her.

Karen's father remained neutral, siding with neither his wife nor his daughter, on these social questions, but when George graduated and went to work for NOAA, instead of join the firm, his interest in Karen's education increased. Genetically engineered soybeans were boosting the yield per acre, permitting the use of herbicides that would kill the natural variety, and adding an uncertain to a market he had come to know well. Perhaps soybeans would soon be drought or freeze resistant. He did not understand the potential or limits to this "genetic engineering" and feared it would transform the market in unexpected ways.

Karen was naturally interested in animals, even the barnacles that she scraped from her boat each fall when it was hauled from the water for winter storage. (It was a Lighting, not a large boat, and she did not race it, so barnacles were removed only annually.) Without any prodding from her father, one fall she brought home some waterlogged wood which was covered with barnacles. She placed small pieces separately in different large jars. She was evaluating their resistance to less salty water, for extra credit in her tenth grade science class. She became fascinated with their delicate feeding lace and some of the tiny worms that emerged from the wood, etc. Yes, Mr. Goldwater said to himself, with a little careful guidance, she could become an expert in genetic engineering.

He did not wait for her birthday. For no apparent reason, he presented his daughter with a high-quality dual-eyepiece microscope. It was immediately used to look at paper, hairs, ants, flies' eyes, and other common objects but soon sat unused on a shelf in her bedroom for the remainder of that school year. In spring, when her Lighting went back into the water, Mr. Gold water was delighted when Karen brought home a plastic bag filled with some seaweed that had clung to the rudder. It had snails and other tiny creatures adhering to it. She examined them closely, but soon the microscope was consigned to the shelf again. His next strategic gift was some petri dishes and growth media that could be used when the Lighting was back in winter storage. As he expected, they remained unused all that summer. Realizing that his overbearing approach with George may have been the reason George did not join the firm after graduation, he was more subtle with Karen. Among her Christmas gifts that winter was the book *Exploring with a Microscope* and the set of microscope slides that came with it, but they were not needed. Karen was already growing bread molds and transferring them selectively to her petri dishes. She had not yet told her father, but she was trying to repeat Fleming's discovery of penicillin for her eleventh-grade biology class.

Mr. Goldwater was pleased with the progress of his less heavy-handed campaign with Karen. There was still a year to go before she would apply to colleges. The horse that Karen usually rode at the riding school died just after Christmas. Shortly after that Karen announced that she wanted to be a veterinarian. A few days later, her father said that Cornell had one of the best veterinary schools in the country. He did not mentioning the fact that it was also developing an outstanding program in genetic engineering.

Mrs. Goldwater was not pleased when Karen's father mentioned Cornell. She wanted Karen to go to school in the Boston area, preferable Radcliffe, where both her grandmothers had gone. By the time the lighting was back in the water the following spring, Karen's interest in becoming a veterinarian had faded, but she frequently brought home samples of water, both from the sea and stagnant ponds, to inspect with her microscope. Both parents were partially successful in this struggle over the selection of Karen's college and career. As the reader already knows, she went to Radcliffe, but she took every possible elective she could in biology, including some at Harvard. (Harvard and Radcliffe have affiliated.) Currently she has several experimental plots of rice and is cross breeding them. After the dark visitor passes, she says it could be important to understand rice. According to George, rice may be the only grain that will grow during the wet summers of South America after the dark visitor has transformed the world's weather.

Chapter 3

▼

Jack

I have told how I met Jack. How Jack met, first George, and then a few months later Karen. That Jack was an astronomy graduate student working towards a Ph.D. in Harvard until his father died and he had to return to the family cattle ranch in South America. The reader knows that Jack returned to Boston when Karen graduated, ostentatiously to select equipment for his observatory, but actually to persuade Karen to marry him. Unfortunately, there is little I can tell about Jack's family. I never met them and both parents are now dead. I know few of the details about his courtship of Karen, except that it was steady and faithful.—He never went out with anyone else. I know that Jack learned to sail with Karen as his teacher. I went out on the bay with them twice, but the Lighting is really a two-person boat. Thus, I will first tell what I can about Jack's activities during the nearly three years we shared an apartment in Cambridge and then tell how Jack first learned of the dark visitor, but did not understand what it was at the time he was first troubled by it.

His time at Harvard and the CfA:

Jacks first year as a graduate student was concerned with course and laboratory work, not research. Although I was already doing research for my thesis, he was not. He wanted to find a thesis problem and an advisor who would accept him. Apparently when astronomers speak of stellar and planetary atmospheres they are concerned with the chemical composition and physical state more than things you and I would call weather (storms, winds and precipitation). By spring of his first year Jack was no longer thinking that he would try to model the red spot of Jupiter as a hurricane. That was a weather problem, not an astronomical problem.

In one of Jack's first year labs he repeated the experiment by which helium was discovered. I am pleased to say that I already knew that helium was discovered by Nobel Prize winning chemist William Ramsay in 1896 by observing the sun. He later went on to discover Neon and Xenon, but that was in the laboratory. (Before I became more focused on the political / legal problems of women in changing societies, I read extensively about the history of science.) Jack showed me the photographic plate he made. It is a long narrow piece of glass, much like a large microscope slide in shape. It did not look very impressive—just a black streak stretching along the length of the glass plate that faded out at both ends. Below this streak and perpendicular to it were a few narrow black lines. They were as long as the streak was wide. Jack explained that the streak was the colors of sunlight spread out by a spectrograph, which initially seemed strange, as the streak was only black or gray at the ends. (Instruments that disperse light into separate wavelengths that can be photographed are called spectrographs.) What Jack meant was the glass plate was a black and white photographic negative made with sunlight after the spectrograph had spread the colors out along the length of the streak. The fine separated black lines below the streak were made in a second exposure of the plate to a "calibration lamp" which gave off only a few very-pure colors of known wavelengths.

One had to look very closely at the black streak to see the point of it all. In the black streak there were a few places where there were fine clear lines across the streak. Jack explained that they were caused by absorption of the sunlight in cold gases in the outer layers of the sun. (I saw several clear lines across the streak. Most were not made by Helium, but one I saw in the "yellow" part of the streak was made by Helium absorption.) The wavelengths of some of the fine clear lines did not correspond to the spectrum of any gas known on earth at the time they were discovered, so this unknown gas was called helium from the Greek, *helios*, meaning sun.

The reason I tell this story is two fold. First I do understand it and secondly, it was this work with spectrographs that re-interested Jack in stellar and planetary atmospheres (in the astronomer's sense of the word atmosphere this time). In ancient times astronomy was a practical aid to sailors and farmers or the priests who use it to tell them when to plant their crops. Astronomy has always had a non-scientific cousin, astrology, which continues in importance today. One can make a plausible case that Hitler lost W.W. II because he consulted astrologers who advised that the stars were not favorable to invade England after the English disaster at Dunkirk.

Astronomy has always been an observational rather than an experimental science. It was the principle subdivision of mathematics more than a thousand years before Kepler and Newton, transformed the mathematics into a form free from the idea that only the most perfect forms (circles and spheres) could be used to describe the heavens. Newton also laid the foundations for modern astronomy when he dispersed sunlight into its component colors with a prism, but he had no photographic film to record the solar spectrum so he did not invent the spectrograph.

The coupling of the spectrograph to the telescope marked the beginning of modern astronomy as a subdivision of physics, although mathematics continues to play a vital role in modeling the data collected. Without this union of telescope and spectrograph we would not know the universe is expanding. This knowledge about the expansion of the universe is based on the "red shift" of lines produced on glass plates and other observation of near-by variable stars. The details are not important for Jack's story, and I would be hard pressed to give them accurately, so I'll return to the story.

Jack thought that there was little prospect he could do anything original in a reasonable time with spectrographic observations of stars. As he put it "All the easy things have been done—it is a very mature field." He is probably not correct in this view. If there is one thing a historian can teach a scientist it is: No matter how mature the field, there is always more to discover, sometimes things that will completely transform our understanding, like Einstein's relativity and the quantum view of physics. I did not try to persuade Jack of this. I was too busy with my own research and sure that there were many problems for him in the field of planetary atmospheres that he was now attracted to for his Ph.D. problem.

Jack considered a lot of possibilities, but just getting an idea that was attractive enough to get observational time on a good telescope was tough. There was so much competition for telescope time. He decided that he needed to try to find something that would not require too much observational time and would not have already been done because the opportunity to do it seldom occurs. Solar eclipses almost satisfy his requirements as they last only a few minutes and are so rare in any one location that most people have never seen a full eclipse of the sun, but they are visible somewhere on earth every few years. No matter where they occur there is always a small army of amateurs and many professionals observing them so Jack gave up trying to find a problem related to solar eclipses. I suggested that he do something theoretical, but he said he was weak on mathematics and did not enjoy "pencil pushing." (Ironically his mathematical success in predicting the dark visitor will make him famous some day.) He was hoping that one day he would have his own observatory high in the Andes and said he needed to learn all he could about observational skills.

In addition to blocking our view of the sun in a solar eclipse, the moon eclipses many other stars every night. (In case you did not know it already, the sun is just a typical star we happen to be near. There are much bigger stars. Some hotter and some cooler.) Mars also eclipses a star or two observable in telescopes nearly every month, more when it is near the earth and less when it is farther away and appears to be smaller. Even more distant planets and major asteroid may occasionally eclipse a star visible in a telescope. With precise timing of when a star is occulted and many observers in different locations on the earth, amateur astronomers make a valuable contribution to knowledge about the shape of asteroids. You could deduce the shape of a traffic sign by moving your head to slightly different locations and noting when the sign hid a distant street light and when it did not. That is the method they use. To encourage amateurs to do this, *Sky and Telescope* magazine publishes in their January issue each year a list of times when known asteroids will occult stars.

Jack initially considered that because of all this existing activity, he was not likely to find a good problem for his Ph.D. related to stars being eclipsed by asteroids or distant planets, but then he remember his first year experiment with a spectrograph and the observation of solar helium. He realized the absorbing gas that selectively removed only narrow wavelengths could be anywhere along the line of sight between a stellar source and his telescope / spectrograph. It did not necessarily need to be near the surface of the star as it was in the case of helium absorption of sunlight. The absorbing gas could be in the atmosphere of a planet with the starlight passing through it just before the star passed behind the solid part of the planet.

Jack was not expecting to discover a new element like helium in the atmosphere of a distant planet, but their atmospheres contain gas mixtures that are not well known and may vary with altitude or distance from the sun, if the planet has a very elliptical orbit (Pluto) or a very tilted rotational axis (Uranus). These factors modulate the surface temperature and can change the atmospheric composition as the planet travels in its orbit. Thus, even if the atmosphere had been observed before by this method, the atmosphere might be better understood if he could measure the absorption of a star's light in a planet's atmosphere just before the star disappeared behind the planet. This seemed to be the type of problem he could use for his Ph.D. It required little observation time and did not happen very often. He might not discover anything new, but taken as a whole, the project could still be worth a Ph.D.

Jack was excited by this idea all morning for one day and then depressed for about an hour that evening after he completed some preliminary calculations. I won't put into print the expletive he used, but it refers to a well-known body function. Even when he assumed the planet was not moving and that the only reason the star disappear behind it was due the fact the earth was moving, the star disappeared behind the planet too quickly. Most of the absorbing gas would be close to the planet's surface as is the earth's atmosphere. Even with the assumption that the planet's gas atmosphere was many times thicker that the earth's, there was just too little time to collect enough light and spread it out in a spectrograph for the photographic plate to develop with any of the telescopes he had access to. I. e. the total exposure time from when the star entered the planet atmosphere until it disappeared behind the planet was too short when he assumed that the earth's orbital speed was "moving" the star behind the planet. Suddenly he said: "Stupid!" followed immediately by "of course." His "Stupid!" was not referring to me, but to himself.

Planets are called planets after the Greek, "*astéres planétai*" or wandering stars. Unlike true stars, they have no fixed position in the heavens but progress all in the same direction about the sun, but when viewed from the earth, the more distant planets appear to pause, turn around and go backwards for a while (known as retrograde motion) and then resume their "prograde motion". For distant planets whose own motion seen for the earth is slow, the retrograde motion is dominant when earth is on the same side of the sun as the planet. Then both earth and planet are going in the same direction but the earth is traveling faster. It is as if you were on a train passing some cars that are traveling on a parallel road in the same direction, but more slowly than the train is traveling. The cars appear to be going backwards as you pass them. When the earth is on the other side of the sun from the planet, it is traveling in the opposite direction from the planet. Then the planet is in prograde motion, but it may not be visible except low in the sky before sunrise or just after dark. When the planet is high in the sky, it is daytime and the planet can not be seen unless you are an astronaut high above the earth's blue sky.

What Jack realized when he called him self stupid, is that twice each year the distant planets are nearly motionless in the heavens as they change from prograde to retrograde motion and approximately six months later when the change is from retrograde to prograde motion. They are never completely still as the turn is a small loop, not a 180 degree reversal. They are almost still when they turn, except for Pluto, as they are nearly in the same plane as the earth. If he were lucky enough to find a bright star in their path near this turning point, then an exposure long enough for the spectrograph plate to develop the long black streak corresponding to the colors of the starlight might be possible. Unfortunately, the star would either (1) be relatively low in the sky around midnight when atmospheric absorption in a long path through the earth's atmosphere would be a problem, or (2) be visible high in the sky near sunrise or sunset when sunlight scattered by the high atmosphere might be a problem. (The planet's slow-motion period and turn occurs roughly when the earth's velocity is directed towards or away from the planet.) His hope was that this unwanted absorption or scattered light could be measured and corrected for.

(Readers not interested in technical information about Jack's Ph.D. problems may want to skip the remaining paragraphs of this subsection, if too technical to interest them. Begin again at the next subsection—"First hint of trouble.")

How he planed to make this correction was clever, so he told me, and I think it is true. He would use what he called an "a-stigmatic spectrograph." That is, one where each point of light on the entrance slit was focused into a single point spot of light on the photographic plate. Actually this is true separately for each color only-the different colors are still spread out in a line, but this line is only as wide as the diameter of the spot of light falling on the entrance slit. It would form a long black line instead of the wider black streak that most stigmatic spectrographs make. The reader should not feel bad if this is hard to understand. Jack explained it to me several times before I "got it." I'll try to make it more clear. (Repetition helped me.)

Most spectrographs have considerable astigmatism. That is, even if only one point of light is on the entrance slit and the entrance slit is illuminated only by a single color from a calibration lamp (Filters can be used to remove the other colors typically produced by the calibration lamp.) a short line is produced on the photographic plate instead of a spot. When white light containing all the colors is used to illuminate the entrance slit of a typical spectrograph with only a single point of light, the wide black streak is formed on the photographic plate by the astigmatism of the ordinary spectrograph. This streak is just many of these short lines side by side as all colors are present in white light.

Jack planned to align the entrance slit of the special spectrograph (His "a-stigmate" he called it.) with the path of the star's arc in the heavens. (I am sure you have seen long exposure photos of stars that appear to be short arcs, all centered on the North Pole Star.) It is these arcs (caused by the earth's rotation) that I am trying to describe. As the star moves along its arc the long black line in the a-stigmatic spectrograph moves parallel to its self. Even if the exposure lasted for many minutes, the arc is so short that it is essentially a straight line and very short. The long black line on the photographic plate in the spectrograph would not move sideways very much during the exposure because the stellar image would not move much on the entrance slit during the exposure. Thus Jacked planned to move the plate during the exposure to enhance the width of the black line recording the colors or wavelengths. The width produced by the star's motion along the entrance slit of the spectrograph would be adjustable. Bright stars could have their light spread out more sideways on the plate. This would permit him to control the exposure of the plate to prevent either over or under exposure. The exposure time could not be changed. It had to correspond to the time the starlight was passing through the depth of the planet's atmosphere.

With Jack's spectrograph design, one side of the photographic plate's long black line of spread out colors (different wavelengths) would be exposed by star light that did not pass through any of the planet's atmosphere. This side of the spectrograph's wide line of wavelengths recorded only the effect of absorption in the earth's atmosphere. This side of the wide line was the first to be exposed. It was exposed while the star was still moving towards the planet. The other side of the wide line was the last to be exposed. It terminated with the disappearance of the star behind the solid surface of the planet (or faded out if the planet was all gas with no solid surface). That edge of the line included the greatest effect of the planet's atmospheric absorption because the rays of sunlight were just grazing the planet's surface and traveling though the longest path possible in the planet's atmosphere. That side of the line also included the effects of absorption in the earth's atmosphere so it was the difference between the two sides of the wide line that was his measure of the absorption by the planet's atmosphere. For most places along the photographic plate (different wavelengths) there would be no differences because the planet's atmosphere was transparent to that wavelength (no absorption). Hence it was only from the few places on the spectrographic plate that the two sides of the black line differ that Jack would get the data he could use for his Ph.D.

The analysis was much more complicated than I have explained (or could). One complexity had to do with the fact that the absorption included on the last edge of the black line exposed measured the planet's atmosphere absorption in all the planet's different atmospheric altitudes. That is, the starlight rays recorded just before the star disappeared passed though the planet's atmosphere near the surface and also though the highest parts of the atmosphere as they entered and left the atmosphere of the planet. The part of the long black line near that final edge also recorded absorption by many different altitudes, but did not have any absorption by the gases nearest to the surface. Likewise, strips of the long black line further away from the final edge recorded absorption from only higher levels of the atmosphere. All of the strips had to be processed separately to eventually "unmix" this mixture of absorption in different layers. Jack told me this unmixing was complicated but not too hard with modern computers. Apparently the programs that do this unmixing already exist.—Jack was not worried about this part. He was more worried about finding the right star and getting time on a telescope when he needed it.

Another complexity had to do with the problem of converting the "clearness" of the line or magnitude of the difference between the two sides of the exposure strip into information about what gas and how much of it there was. Jack did not even try to explain how this was to be done and I don't know how. It seems to be an impossible task to me, but Jack said he could do it.

Jack's plan for aligning the motion of the star's image along the length of the entrance slit had another advantage that was of some importance in this regard. The telescope did not need to track the star automatically with high precision. With the entrance slit of the spectrograph aligned with the arc of the star's motion, the tracking could be manual. Most telescopes, even some small ones sold to amateurs, have precision motor drives that compensate for the earth's rotation. I.e. track the star's motion in the heavens, but only very large telescopes would tolerate the extra weight of an attached spectrograph when tracking a star. The demand for observational time with one of these large telescopes is so intense that he could wait years to get time on one big enough to carry the weight of his small spectrograph and still track the star. He could attach his spectrograph to an older, more modest size telescope (but still a large one to collect the light he needed) without unbalancing it and do any tracking required manually. Most observatory sites have several older telescopes of modest size, still in use, or at least easily restorable for use. These older units do not have such long waiting lists of users. He could use one even if its star following drive was broken or being repaired.

At the time Jack first thought this would be a good Ph.D. problem he did not realize how rarely distant planets occult stars. He ended up working mainly with Mars, not a distant planet for this reason. I have explained his plan with reference to photographic plates, because this is the way Jack explained his ideas to me, but in fact Jack's final design used a "linear array CCD," which is something like the "film" of a digital camera, instead of a photographic plate. The CCD is much more sensitive and responds in the infrared where molecular absorption takes place. (The atmosphere of Mars has a lot of carbon dioxide molecules.) There was also a possibility of making a measurement or two with some of the giant gas planets, but tiny and distant Pluto was not expected to occult a significant star until June of 2008 and even that star is seven magnitudes dimmer than the faintest star visible to the human eye! (Because of the disturbance of the dark visitor, this predicted event will not happen but others will, especially when Pluto is closer to the earth.)

Jack spent the second half of his second graduate year designing his spectrograph (actually modifying the optics of an existing one and making it lighter). He was also busy designing the "light tight", but movable glass plate holder. In addition to this design work, he still had some course work to complete. He took his final qualifying exams that spring and found a thesis advisor who had both some funds to get the photographic plate transport made and enough influence with telescope time allocation committees to get Jack time when he needed it on one of their smaller telescope. (His Harvard University thesis advisor held a joint appointment at CfA and this helped.) Jack had constructed a list of times when various planets would occult suitable stars. He wanted to try to observer them all.

He had made only one partially successful measurement when his father died and he had to return home. He arranged for the plate transport and modified spectrograph to be stored, telling his professor that he would try to return to complete his Ph.D. research and flew home for the funeral. In South America, he wrote up the observation he had made, the design of his instruments, and the details of his analysis plan. Based on this he received a Masters Degree in absentia the following spring, one week before he returned to see Karen graduate. While he was purchasing his mirror and other things needed for his observatory, he persuaded his professor to let him take the modified spectrograph and plate transport home, as an indefinite loan. Both belonged to Harvard, but no one was planning to use them. Later, Jack told me that if he had fully understood how tough his problem was when he started; he never would have tried to use them either. Because of these difficulties and the absorbing demands of his efforts to understand the dark visitor, this equipment and Ph.D. project have been abandoned, at least for the present. After the dark visitor has passed the sun, Pluto should have more atmosphere because it will be nearer the sun and warmer. Perhaps Jack will eventually complete his Ph.D. project, if not with Pluto, with another planet or with one of the many newly evaporating comets that will be thrown into a highly elliptical orbit by the dark visitor. Jack is very smart. I would be willing to bet he eventually gets his Ph.D. At least he will be the world's most famous astronomer, soon.

First hint of trouble:

During the last year of his time at Harvard, Jack adopted a strange sleeping pattern. He would spend most of the night at Harvard / CfA facilities and return home to go to bed just as I was getting up to eat breakfast. We ate our cereal together and then he went to bed for "one sleep cycle," approximately four hours. After diner, which I often cooked in the apartment, he returned to bed for a second sleep cycle that usually lasted only three hours. By 10 PM he was on his way back to the CfA, hoping someone would be doing something interesting or better still have a problem and let him use the remainder of their allotted time. (Some of the telescopes are in better seeing areas and remotely controlled. If nothing else, he spent the time examining the extremely large collection of previously exposed plates.) He claimed that the US navy had demonstrated that two separate sleep cycles were better than one continuous night of sleep and it seem to work for him.

A few months before going home to manage the cattle ranch, Jack decided to photograph Pluto and its moon, Charon, in different positions whenever there was time available. In addition to his own observations, he also collected all the old records he could find at CfA, but he wanted his own photographs to display in his observatory. Charon is about half as large as Pluto is and they circle about their common mass center in less than seven days. They are separated by less than nine diameters of Pluto. Until recently Pluto was closer to the sun than Neptune, but it is now again the most distant planet and moving farther away each year. Because of its high orbit inclination and location (now and for the next 150+ years) Pluto is easier to observe in the Southern Hemisphere. Jack planned to make a life-long study of Pluto's motion, accurately measuring its position and looking for perturbations. The mirror he bought can just resolve the disk of Pluto and may be able to do this for Charon also, when seeing conditions are exceptionally good.

He planned to eventually decorate the walls of his observatory with photos of all the planets. He thought it would be wise to start with Pluto because it would never be so close again in his lifetime. He planned to make a final composite photo showing both Pluto and Charon in about five or six different positions for a wall decoration. He took six or seven photos of the pair spread over several months at CfA but had not been able to complete the set he wanted when his father died. The ranch foreman made all the funeral arrangements and Jack arrived home the morning of the funeral. They buried his father next to his mother in the afternoon shade of the tree his grandmother had planted.

The reader already knows that he stayed in South America to manage the cattle ranch and settle his father's accounts etc. He only returned to the USA the week after he had received his Masters degree, just in time to see Karen graduate. After they married, both returned to South America and within one year of their return, his observatory was functional. After initial collimation and alignment studies with his telescope, he began again the series of Pluto / Charon photos in South America. (This was a good test of his telescope and no stars were scheduled to be occulted by distant planets for several months. Also he had lost interest in returning to Harvard to complete his Ph.D.) He planned to begin a study of Mars, because it also is best seen from the Southern Hemisphere about every 26 months but it was too soon to start.

One of the items Jack had imported for his observatory was a backlighted digitization table, used to measure the positions of star images formed on glass photographic plates. With it you could easily measure the exact position of a comet or planet against the background stars and then feed the results into a computer program that would automatically calculate the orbital elements (mathematical description of the orbit) from three or more observations separated in time. Jack want to test it and already knew what the orbital elements of the center of mass of the Pluto/Charon system were so he use both the series of photos he took while still in the USA and his new series to compute the orbital elements. The results for the two series almost agreed, but the error was more than it should be. At first he assumed that atmospheric refraction in the earth's atmosphere or the parallax difference between the two observation sites could explain the disagreement, but rapidly discarded this idea for two reasons. First the computer program he used had inputs corresponding to the observatory location, even its altitude, and secondly the Earth's movement though space during the period between separate measurements, which was also corrected for by the program, was much larger than his move to the Southern Hemisphere.

He thought the most likely origin of the problem was a small defect in his digitization table. Jack also had several glass-plate photos of asteroid Ceres he had made at Harvard / CfA. When he used the digitization table with them, the computer produced the accepted orbital elements. (This is why he discarded the atmospheric refraction hypothesis.) After a week of pondering what could be wrong, Jack decided that there was nothing wrong with his digitization table or computer program. A large asteroid must have hit Pluto between the time of the USA and South American measurements.

Even though mathematics was not Jack's forte, it was less than one day before he knew the momentum that the asteroid must have transferred when it hit. He was half way though writing a brief note to Email to the CfA's Minor Planet Center when he realized his observations could not be explained by an asteroid collision unless the rotational period of the Pluto / Charon system was also changed. (The asteroid would transfer angular momentum as well as linear momentum to the system.) Returning to the digitization table, he began to use both series of photos to carefully evaluate the rotational period of the system. There was no change. Both sets gave a period of 6.387 days. Any change in the period was less than five minutes and this was not consistent with the linear momentum he was about to report to the CfA / MPC.

The more he tried to build a consist understanding of what was wrong, the less he understood. Something had slightly changed the orbital elements of the center of mass of the Pluto/Charon system, but had not effected either their separation or their rotational period. What had started as a simple wall decoration project was now keeping him awake night and day. He continued to make careful measurements of the Pluto / Charon system for more than a year before he contacted me again. He can't remember exactly when during this period he thought of the dark-visitor explanation, but he thinks it took about two months after doing so before this idea was firm in his mind as the best possible explanation.

Thus, there is no definite date on which he thought of the dark visitor idea or when he concluded that it was the only consistent explanation for all of his observations and the earlier data. Initially, the dark visitor was just "a gravitational disturbance by a massive object traveling towards the solar system with approximately the speed of the local group" (Our galaxy, the Andromeda galaxy and a few others). Jack had no idea where it was. He only knew that it had to be far away so the both Pluto and Charon would experience essentially the same gravitational field and not move relative to each other or change their rotational period. If it was far away, it had to be much more massive than an asteroid. It had to have at least the mass of the sun. If it had the mass of the sun and was approaching our solar system why had no one observed and reported it?

Thousands of amateurs scan the heavens looking for comets every night. The thing that seemed to make the most sense was that as it was a small black hole (no reflection of sunlight) that had moved through space with the local group but was now approaching the sun. Thus, it had probably formed in the local group and had slowly drifted towards our sun for a hundreds of thousands years. It was only by chance that the sun was now near its path, but soon they would start to "fall" towards each other.

The next few paragraphs are a summary of Jack's teachings (presented in detail in Chapter 8) about all the things the dark visitor might be. I include it here for the less scientifically inclined, who may not be interested in the details of stellar evolution and the death details of our sun (and of the earth). They may decide to skip Chapter 8. Doing this causes some redundancy so the scientifically well-versed reader may want to skip to the next chapter now and wait for Chapter 8 to read Jack's more complete version. I encourage most readers to read both as this simplified version may help them understand Jack's more complete version.

When massive stars die their core is iron and even iron can not support the force of its own weight. The core collapses into at least nuclear matter density in a violent event that releases so much gravitational energy that the outer layers of the star are blown away in an event we call a supernova. Initially the outer layers of the dying star are too dense for the released energy to pass through, so a shock wave forms and this is what blows them away. The best known example of this is the Crab Nebula supernova explosion. This shock wave is far stronger than any we can ever produce on earth. So strong that it drives nuclear fusion more powerfully than a hydrogen bomb. All the elements heavier than iron that exist in the universe were made by these shock waves. Life could not have arisen on planets circling the first generation of stars created after the "big bang" unless in some form that did not require elements more complex than iron. Our sun is at least a second-generation star as it has elements more heavy than iron in it in small quantities.

Astrophysicist have shown that the core that remains after a star "dies" will normally be a "white dwarf" or a "neutron star" or a "black hole." If the core mass is less than 1.4 times that of the sun, it will be a white dwarf, if more than about 2.5 solar masses it probably will be a black hole, if in between these two limits it is likely to become a neutron star. The Crab Nebula explosion left a neutron star behind.

Neutron stars are also usually, if not always, "pulsars" in that they emit brief pulses of radio waves at extremely regular intervals. I know about this because I have always been interested in the history of science. These regular radio pulses from the heavens were observed before their neutron star origin was understood. Initially these regular pulses were taken as evidence of life in other solar systems. Pulsars were originally called LGMs, which stood for "Little Green Men," even in some of the scientific literature.

I mention this, despite it having nothing to do with Jack and the dark-visitor story, just to show how interesting history can be and also to note the important role women have played in science ever since the time of Marie Curie. (A woman discovered the first pulsar.) History is not only interesting, but also extremely important for policy decisions that effect millions. George Santayana said it best 100 years ago: "Those who can not remember the past are condemned to repeat it." Political leaders resisting full legal rights for women, should study the economic advance the Scandinavian countries achieved when they granted full legal rights to women long before the suffragettes were marching in the USA. I apologize to the reader for this one paragraph tirade, but it is something I feel strongly about so I have not removed it. Now to the stars.

Any white dwarf near enough to have disturbed the Pluto / Charon center of mass trajectory would be so bright that it would be very visible during the day. Thus Jack knew that the distant massive object was not a white dwarf. The typical neutron star was also ruled out as even if were a very weak pulsar, it would have been detected long ago. (Pulsars emit energy in all parts of the electromagnet spectrum, even X-rays.) He was not willing to rule out anything else theoretical possible. He thought of several alternatives, including an assembly of magnetic monopoles, which have never been observed, but are theoretically possible.

Jack is of the opinion that the dark visitor is a black hole, probably the residual core of a supernova star pair in our galaxy. He thinks the first member of the pair passed very distant from the sun during the 1920s and the second is now approaching. His initial estimate was that the black hole now approaching would have several solar masses, as these are the most common black holes. Very much more massive black holes exist at the center of most, if not all galaxies, but the dark visitor is likely to be a simple, common black hole left over from the death of a large, previous-generation, star (not from an earlier generation, or from a distant part of our galaxy, for reasons Jack explains in Chapter 8.) The dark visitor can't be a large black hole because it would then need to be many light years away to produce so little disturbance to the Pluto / Charon system and if that massive, it would have had other effects on near by stars. Jack is quite willing to accept that it is something else, perhaps the dark visitor is made of the mysterious "dark matter" that is now believed to constitute more than 90% of all the matter in the universe.—See Chapter 8. Whatever it is, it is coming our way and will forever transform life on earth.

Chapter 4

▼

Amiel

After he sold his younger brother to the Chinese pharmacist, Amiel went straight to the harbor. He quickly found the Star of India, the ship the old man had recommended to him. He asked for the first mate by name and, just as the Chinaman had promised, the first mate took him on as a deck hand. Amiel had never been to sea before but he was strong and quick to learn. They set sail for Bombay two days later.

During his walk to the harbor, Amiel had wondered if everything was foreordained, perhaps even foretold.—Why else was his brother named Joseph?—He felt guilty, but he had had no other choice. All they had eaten in the last four days was two small loaves of stolen bread and a few onions they had gleaned from a field near Copenhagen. Amiel told himself that when he could, he would return and buy back his brother's freedom, but he still felt guilty. Many years later, a residue of that guilt may have killed him, as the reader will soon learn.

The Chinaman had told Amiel that the first mate owed him a "favor" and would give Amiel meals and a berth, but by the end of the first week at sea, Amiel was sure that the favor went the other way. He learned from other members of the crew that he was not the first deck hand the Chinaman had tricked into signing on for the long haul to India. The Chinaman often supplied naive deck hands to first mates of ships bound for China and India in exchange for herbs and spices. Conditions on the Star of India were terrible. Amiel had scarcely more to eat now than when he was on land. When he complained, he was flogged—six lashes—and told that if it happen again, he would get the full dozen. After that he keep his mouth shut, but when the ship put into Cape Town Harbor for fresh water and supplies about a month later, Amiel waited for night and then lowered himself silently into the water and swam to shore. He nearly drowned. The shore was much farther away that night than it had been earlier in the day when the long boats were returning with supplies.

Life on land was not much better. The Boers had little use for a Russian Jew. One month later he was back at sea, bound for Adelaide. After that he lost track of the ports. It was about ten years before he got back to Copenhagen. He found the building where the Chinaman's pharmacy had been but it now sold dry goods, including silks from the orient and carved jade. The owner was again an old Chinese man. He told Amiel how his brother had taken a wife who died a few months before he bought the store. He was not sure, but he thought his brother had gone to America after selling the store. Amiel was relieved to learn that Joseph, like his Biblical namesake, had prospered during his years of servitude. Amiel himself had nothing to show for his years of often-cruel labor. When he left the shop the old Chinaman gave Amiel a small jade ring, saying he had found it upstairs and thought it might have belonged to his brother's wife. Amiel was hungry and that ring bought him the best meal he had had since his father had killed the family's goat, more than a dozen years earlier.

Refreshed by that meal and relieved of the guilt that had brought him back to Copenhagen, he decided to return to Russia and try to find his father. A few days later, he signed on to a coastal steamer, bound first for Helsinki and then for St. Petersburg, but the steam engine failed just off the German coast. Fortunately the converted ship still routinely used her masts and sails when there was a good wind to save coal, so they put into Hamburg for repairs and cheap German coal with little trouble. These new steam motors were not very reliable but he had baked in the subtropical sun too many times without wind not to recognize that they had potential. The repair yard was busy and they had to wait their turn.

The yard was also short handed and he needed money so he decided to help out while waiting for the ship to sail again. Then he heard about the new pogroms and decided he had had enough of the sea. So when the ship finally resumed its journey to Helsinki, Amiel stayed behind, still working in the yard. He rapidly learned about steam engines, and developed a better way to remove accumulated boiler scale. A few years before the turn of the century, he was in charge of the yard, married and had two "kinder" by his German wife. He was now too old to do much of the hard labor himself, but he told the younger men what to do and he commanded the respect of all. He even occasionally went to a Lutheran Church with his wife and kids, but he no longer believed in any God.

One day three German naval officers came to the yard and showed him some design information about a compact diesel engine. They wanted the yard to mount it in a section of steel tube they would supply and to test it four meters under water with the exhaust being discharged under water also! Air for the engine would come down from the surface in a long pipe that stuck out of the water. They wanted the engine mounted on springs and the exhaust had to be discharged below the water as quietly as possible without back-flooding the engine when it was turned off. He was not to ask any questions as to why or tell others what he was doing. Just give different workers their instructions as needed and nothing more. If he made it work quietly under four meters of water, he would get a big bonus and German citizenship. The yard would also work exclusively for the German Navy and have many more profitable contracts. Some sailors would be assigned to the yard to keep it secure. It was as we now say "an offer he could not refuse." He could be deported to Russia. No threat was necessary. Amiel looked forward to this new job and the bonus it offered. He had never been able to save much money, and he was already 63 years old. He wanted to retire. If all went well with the German Navy, he could afford to do so in a few years.

When he married in his early forties, Amiel had again tried to contact his long lost brother. Amiel got the name of an orthodox rabbi in New York City and wrote to him, giving all the information he had learned from the second Chinese owner of the building where he had abandoned his brother and a description, even though his brother was only 13 when he last saw him. Eventually another rabbi wrote back that he had known a Russian Jew who had lived in Copenhagen before coming to New York. That man resembled Amiel's description of his brother, but his name was Gulvassor. He had married the niece of another rabbi, now dead, and gone off to Boston with her to live. He was sorry that he could not help Amiel located his brother.

A year later, an American sailor from a ship waiting to be repaired was working in the yard as Amiel had done several years earlier. This sailor had been in Boston a few years back and had met a man in a waterfront bar who looked very much like Amiel. He did not remember the man's name but it was something Jewish or foreign sounding—something like Amiel's own name. That man hated his wife and still loved the young Jewish girl he had married years earlier in Copenhagen. She had died trying to bear his child. When Amiel heard this, he was sure the New York rabbi had remembered his brother's name incorrectly.

He again began to try to contact Joseph via a Boston rabbi. Once again he missed his brother by a few years. His brother had been dead for three years by the time another Boston rabbi replied. That rabbi was new to the area and had never met Amiel's brother. Amiel had learned to write some German from his wife but he had a friend write the first letter in English for him. Somehow in all the confusion, the new rabbi thought the second Mr. Gildwaser, Joseph Junior or "JJ" as he was sometimes called, was Amiel's brother instead of his nephew. This confusion was quickly corrected. Uncle and nephew began to correspond, telling each other what they knew about the first Mr. Gildwaser, Amiel's brother and his father.

The letters that Amiel sent to JJ, the second Mr. Gildwaser, still exist but those that JJ sent to his uncle are lost. After the first one, the letters appear to have been written by his German wife. The last one was written in 1912 by the same hand that wrote all the others in German. It tells of Amiel's death in the yard by a firing squad of navy sailors, but his wife did not know his age. She thought he was about 70, but he was at least 75 when he was executed. These letters were found tied together in the office safe by Kathleen in 1952. Karen kept them after her mother died, but she does not read German. I can read German so there was no need to seek help from a translator this time. Most letters already have partial English translations written between the lines or in their margins, but several do not. I read them to her, paraphrasing the German into English.

From the one side of the correspondence that I have read, it is clear that Karen's grandfather learned more than he should have about the German Navy's U-boat development program from his uncle. This is the only thing of any historical significance in the letters. Most of the personnel facts reported in this chapter also come from those letters, but I have invented some of the minor details. To what extent these letters influenced her grandfather's decision to sell his mother's ships I can not tell, but I am sure that Amiel's loose tongue about the yard's U-boat activities was intentional—perhaps a repayment to the second Mr. Gildwaser for selling his father to the Chinaman. Clearly he knew what he was doing.—All that still have envelopes were mailed from Denmark, not Germany. An early one instructs JJ to never mention the U-boats in any reply.

Chapter 5

Climate BDV

First let me acknowledge that all of the ideas and many of the words presented in this and the next chapter come from George. I have, however, edited them extensively and added parts to George's draft whenever I thought more explanation was required. George gets credit for these additions also in that I am only expressing (I hope correctly.) the concepts he explained to me over the phone in response to my inquires for clarifications. (We are both too busy just now to get together for pleasure and beers.)

Climate is the earth's attempt to achieve energy equity (natural justice in action) for different latitudes. Weather is the short-term and local means by which this objective is pursued. Although some weather phenomena will be mentioned, this chapter is concerned with the basic climatic conditions of earth Before Dark Visitor, **BDV**. The next chapter is concerned with the climate After the Dark Visitor has passed, **ADV**. I think I am justified in introducing the designations BDV and ADV, because historically any well-defined event, which transforms the nature of life, is often taken as year one of a new calendar. The Dark Visitor passing the earth is such and event and we will certainly need a new calendar when the year is 378 days long.

The table on the adjoining page illustrates the extent to which different latitudes receive solar energy. I present data for three days and four locations. The columns labeled Max and Min Days correspond to the days with maximum and minimum solar input, for that latitude. The final column labeled Equinox is for either of the two days annually when the sun is over the equator (day and night of equal length for all cities on earth).

Noon Solar Intensity on High Horizontal Surface

Site near	(Latitude)	Max Day	Min Day	Equinox
Havana	(23.5°)	100%	68.2%	91.7%
Boston	(42.5°)	94.6%	40.7%	73.7%
Oslo	(60.5°)	79.9%	10.5%	49.2%
North Pole	(90.0°)	39.9%	0%	0%

The table gives solar energy[2] incident upon a unit of area at noon when that area is oriented horizontal and high above the surface. That is, the values in the table are the percent of the full solar intensity incident upon a horizontal unit of area above the atmosphere of a spherical earth at local noon. Energy that may be reflected back into space by lower altitude clouds etc. or absorbed high in the atmosphere is included. All solar energy is included, not just the visible light. The cloud cover and length of the day are also important factors controlling the climate, which I have ignored here as my intent is only to give some quantitative support to illustrate the idea that high latitudes do not share equally in the sun's gifts. (The "full solar intensity" is approximately 1400 Watts per square meter.)

Climate is nature's attempt to correct this latitude inequality. The fact that a typical mid-winter day in Oslo is often warmer than Boston shows how well nature's climatic efforts have succeeded in redressing this unjust distribution of solar energy. In this particular case, her major tools that keep Oslo relatively warmer are the Gulf Stream and, strange as it may initially seem, snow. In order to understand ADV climate, one must have some understanding of the way BDV climate works. That is the objective of this chapter.

Because it is more familiar to North Americans and European readers, I will focus on the North Atlantic, both ocean and atmospheric circulation, but the same general ideas apply to the Pacific and the Southern Hemisphere's ocean regions. Weather is much too complex for human understanding. Climate is less complex, but I will start the explanation with a simplified earth, the way George did for me.

Imagine that the sun really did go around the earth, just as the ancients believed. That is, imagine that the earth did not rotate. How would climate move the excess equatorial heat towards the poles? The answer is basically the same for

---

[2] The second zero in the table for the North Pole location is technically correct, but some sunlight is present at the ice covered surface due to scattering off of clouds.

both the atmosphere and the sea, but because there is salt in the sea, the sea is a little more complicated. Consequently, I will begin with the air.

Atmospheric circulation:

Everyone has seen smoke rising and most have seen hot air balloons, at least on TV. Air expands when it is heated and then rises. Some colder air must replace it near the surface. The colder air could come from under a near-by cloud. (The sun is not heating this air in the cloud's shadow, so it is colder.) This cooler air could descend to replace the rising warm air (and probably would), but that is local weather, not climate. From a climate point of view, the colder air comes from a latitude closer to the poles. Recall the general latitude effect illustrated in the solar-intensity table that I presented at the start of this chapter. Thus, in the climate of a non-spinning earth, air near the surface would be moving towards the equator and gaining heat as it does so. The higher altitude air would be moving towards the poles and losing heat. That is, the prevailing surface winds would come from the North in the Northern Hemisphere if the earth did not spin.

Before I let the earth spin and thus explain why the surface winds in mid-latitude come out of the West and not the North, let me make one important observation that helps to explain why the air in snow-covered Oslo is not very cold in winter and why hurricanes carry so much energy. It is difficult for the average person to understand how much heat rain or snow can deliver because winter rains and snow are cold. As warm moist air rises it expands and this expansion causes it to become colder. As it cools, it capacity to hold water vapor decreases. When the relative humidity becomes 100% the water vapor changes back to water (or ice crystals) assuming there are enough microscopic particles to help initiate this condensation / freezing. This heats the air more and permits it to rise more. This "heat of condensation" is the same as the heat that was required to evaporate the water initially. If the condensed water freezes, it releases even more heat into the air. Consider how much heat is required to boil (evaporate) a pan of water and then imagine how much heat is released into the air when tons of water vapor condense as rain or snow.

The next paragraph gives a brief look ahead to the outcome of chapters 5 and 6 to encourage the less scientifically inclined reader to struggle through these two important chapters to understand why the ice age will return to the Northern Hemisphere, but not to the southern one. All prior ice ages have effected both hemispheres simultaneously, but the one the dark visitor will cause is fundamentally different.

Most Northerners know that the snows that come in the coldest part of winter are typically fine and dry. It is in the late winter or early spring (when it is not very cold—air temperature near freezing) that the really heavy "spring snows" fall. The milder winters of the Northern Hemisphere caused by the dark visitor will have many of these great wet snows falls every winter after the dark visitor passes. The Southern Hemisphere, sweltering in summer heat while the North is in winter, will supply the water for the North's snows by evaporating ocean water. Much of this evaporated water will be stored on land as ice and snow in the Northern Hemisphere. This is because some of the snow that falls in winter will not melt during the following cold summer when, six months later, the earth is farther from the sun in the new elliptical orbit the dark visitor has given it. Each year more snow will accumulate. The lower layers will change into glacier ice under the weight of the higher layers. In about one hundred years, the ocean level will drop at least 200 feet as ice accumulates on land. Most of America's continental shelf will be ice-covered land above the new sea level. Atlantic City and other beach-front property in the USA will be miles inland from the ocean, but this will not concern the owners of this property because ice will have buried all of the gambling casinos and "beach-front" houses. Now let's let the earth spin to continue with the scientific explanation in greater detail so you will understand how and why all this will happen.

At the equator, both the land and the air are moving at about 1000 miles per hour, if seen from the moon (or any other point traveling through space with the earth). This is because at the equator it is approximately 24,000 miles around the earth and the air must go around in 24 hours. Citizens of Ecuador don't notice this speed because they are also going around at 1000 miles per hour. This "free speed" is why all space-rocket launch-centers are as near to the equator as the launching nation can place them. Relative to the land, the speed of the moving air (we call it wind) is much less, but these high equatorial speeds have an enormous effect upon the climate.

At Lisbon's latitude the distance around the earth is only about 3/4 the circumference of the equator, so there the air is only moving at 750 miles per hour to go around with Lisbon in 24 hours. If you could instantly move a big chunk of this Lisbon air to the equator, the earth would be running under it so fast that the wind would come out of the East at 250 miles per hour. Obviously a mass of Lisbon air can not be transported instantly but recall from the non-spinning earth's climate that the surface winds should tend to come from high latitudes and move towards the equator. Thus as Lisbon's air moves south, it tends to head west also as it is not going around the earth fast enough to keep up with the land (or ocean) spinning under it. By the time this air from Lisbon gets near the equator, it is mainly moving towards the West as seen from the deck of a ship in the tropical ocean. On its trip southward, this air mass gained most of equatorial speed from friction with the surface—making steady winds towards the west, not 250 mile / hour gales!

Local weather phenomena such as warm clouds rising and cold air descending tend to keep the lower parts of the tropical atmosphere mixed and all moving towards the west, at least as the air mass moves west from Africa, but recall from the non-spinning earth model, this warmer air must also tend to go towards the poles to replace the air which is going toward the equator. Thus all year long east of the Caribbean region there is a general climatic circulation of air along the path of typical hurricanes—westward from Africa but turning northward.

Columbus may not have understood why (as I hope you do now) but he knew you could find good reliable winds to the west if you first went south from Spain/Portugal along the West Coast of Africa towards the equator. Even today these dependable winds from the east in the tropics are call the "trade winds" because they were so important for trade between the new and old worlds when ships were moved only by the wind. As surface air that was once near Lisbon or Oslo travels west from Africa these trade winds acquire both heat and moisture from the ocean. This moisture (water vapor) represents a tremendous store of energy, which as the weather begins to cool in the fall, can condense (rain) and release this energy in the form of a hurricane. In the winter, even more energy (heat) can be gently released as this moisture condenses into snow. This helps to keep Oslo etc. from becoming too cold. Even when you understand it, it is hard to believe, but true, that lots of snow helps keeps Oslo warmer than Boston!

In summary, one can say that the excess heat of the tropics and the spin of the earth combine to give a clockwise circulation (when view from space) to the climatic wind patterns in the Northern Hemisphere. Or more generally, for both hemispheres, it is true that the prevailing winds over the oceans in the tropics move towards the west and then turn towards the poles. Thus in the Southern Hemisphere the climatic circulation is counter clockwise.

The clockwise circulation in the Northern Hemisphere is also caused by the fact that by the time this tropical air mass is at Boston's latitude, it is now traveling too fast towards the east for the surface under it, making winds to the east from the west. The same is true in the Southern Hemisphere but it is a counter clockwise circulation that takes air from Argentina eastward towards Southern Africa. The "pole seeking" equatorial air is moving too fast towards the east as it travels toward the poles. (Just as the "equator seeking" mass of polar air from Lisbon or Oslo was moving too slow as it went south.) Thus, at America's latitude, the prevailing winds come out of the west and blow towards the east because they still "remember" part of their equatorial speed. The polar air masses are given a general circumpolar circulation (clockwise in the Southern Polar region and counter clockwise at high Northern latitudes, when view from above the pole) to a large extent by their contact with these tropical / mid-latitude circulations. The relatively faster, high-altitude "jet stream" is more complex and need not concern us. As it wanders, it effects weather more than climate. These colder polar air masses tend to slip under the warmer circulation patterns. Americans speak of a cold Canadian air mass moving in when this happens, but taken as a whole, there is not a great deal of mixing of the air masses either between polar and mid latitude air masses or between hemispheres. The latter fact is one of the things that will change ADV.

Oceanic circulation:

The "force" that causes north or south moving masses of air or water to be deflected sideways (the clockwise turning in the Northern Hemisphere just discussed) is called the Coriolis force, after the French engineer Gaspard Coriolis who explained to Napoleon why his long-range cannon fire was always falling a little to the right of the target if the target was north or south of the gun, but was "on target" if the gun was aimed either east or west. (Napoleon began his career as an artillery officer and was always interested in science[3].) The Coriolis force is not a true force, only the effect of a spinning spherical earth.

The Coriolis force is proportional to the speed of the moving mass. Because the cold polar water flowing towards the equator along the ocean floor is moving much more slowly than the cold polar air, the Coriolis force has less effect upon it. That is, friction with the ocean floor has more time to act and tends to impart the local circumferential speed of the spinning earth to the slow moving water. This bottom flow is more of a gentle drift to the equator rather than a current. Thus, unlike the trade winds, some of the drift of cold bottom water from the north has not turned west by the time it reaches the equator. A significant part of it continues into the Southern Hemisphere. An equal mass of warm water from the Southern Hemisphere drifts across the equator to compensate, but this flow is near the surface. The basic reason why cold water migrates to the Southern

---

[3] Perhaps he intuitively understood the germ theory decades before either Pasteur or Semmelweis. He supported the development of canned food. It can be argued plausibly that canned food was more responsible for his military victories than his generalship. (He could mount long campaigns in distant lands and yet feed his army, but his supply of canned food was no match for the Russian winter.) As Caesar said: "An army travels on its stomach." Late in W.W. II, some German soldiers had no belts—they had boiled and eaten them. The same may have been true for a few US soldiers when Patton was advancing too rapidly for the supply line to keep up.

Hemisphere and warm water migrates to the northern one (an asymmetry) is the two polar regions differ. Antarctica is an ice-covered landmass but the Arctic is an ice-covered ocean, which can flow. ADV, these flows in the Atlantic Ocean will be much larger for reasons I will try to explain in the next chapter. After the dark visitor has passed, the Southern Hemisphere will transfer enormously greater quantities of heat (both as warm water and as atmospheric moisture) to the Northern Hemisphere during the northern winter.

The volume change of seawater with thermal expansion is much less than in air so the density of seawater depends upon its salt content as well as its temperature. These two factors tend to oppose each other because as seawater is warmed, the rate of evaporation is increased. This causes the salinity to increase. The density change of seawater, slowly warmed in an open container, is less than if fresh water were warmed the same amount. The water of the Mediterranean Sea is warmer than that of the ocean but as the surface evaporation exceeds the rain fall and fresh water input of rivers, it is also saltier than the ocean. During periods of glaciation, the sea level falls as more and more evaporated seawater is stored on land in the form of ice. During the prior ice ages, the now flooded continental shelf was exposed land. The Mediterranean Sea was an inland lake that dried up leaving large salt deposits on the floor.

The surface layers of the tropical oceans also become warm and more salty with solar evaporation, but they do not sink to the bottom while they are concentrating salt because of the solar heat keeps them warm. Like the air above them, they move towards the west, but much more slowly. In the Pacific Ocean this movement is mainly due to contact with the surface winds. In the Atlantic it is more complex. As this movement approaches the western regions of the ocean, it turns towards the pole. Partially because land is blocking further westward movement and partially because of the Coriolis force. This gives rise to an enormous "river" of warm salty water traveling up the East coast of America (the Gulf Stream) and the East coast of Japan and continental Asia (the Kuroshio current).

The Kuroshio current is much weaker for reasons that will become more clear near the end of this chapter when the differences between Pacific and Atlantic circulation is explained. For the moment, let me just state that the North Pacific Ocean water is saltier than the North Atlantic because it receives much less of the relative fresh water found in the Arctic ocean where fresh water ice melts during the summer. When the salty Gulf Stream water is north of Boston, it has cooled some and begins to sink under the weight of its higher than ambient water salt content. As it sinks it mixes with the colder bottom water, which is less salty because some of the bottom water is from the Arctic Ocean. This lowers its salt content, but it has already acted as a piston while sinking to push bottom water towards the south. The Arctic water leaving the Arctic Ocean along the bottom must be replaced so some of the warmer surface remnants of the Gulf Stream flow into the Arctic Ocean off the Norwegian coast. This flow keeps many of Norway's ports ice-free all winter. In the North Pacific, the bottom water is already quite salty and the sinking is much slower with less "piston effect."

Because of the Coriolis force, the cold water from the Polar Regions flows towards the equator along the bottom of the ocean mainly on the eastern sides of the ocean basins. (The Coriolis force is weak, but not without effect.) In the North Atlantic as it passes off the Straight of Gibraltar, some enters into the Mediterranean Sea, and an equal mass of warm salty Mediterranean water flows out into the ocean in the surface layers. As this Mediterranean water penetrates into the colder ocean, it loses heat. Because of its high salt content it begins to sink in a self-accelerating process. The sinking part encounters even colder water, cools more rapidly and sinks faster. "Fingers" of "falling" saltier water develop and mix with the cold southbound drift towards the equator. This development of "falling fingers" is a natural example of the "Taylor instability" with a dense fluid unstably posed over a lighter one. These falling fingers provide a second but much smaller "piston effect." Their extra load of salt also helps to keep this south-bound flow along the bottom by increasing its density (salt content).

In the Pacific Ocean the cold deep water flows more slowly off the California coast and does not receive this gift of salt. Before it approaches the equator, it has warmed and begins to rises towards the surface, mixing as it does so. It is rising to replace the surface water the winds have blown westward, not because it is less dense that the surface water. That is, this bottom flow is rising mainly in response to hydrostatic pressure. As it begins again its journey across the tropical Pacific, it increases both its temperature and its salt content. In both oceans of the Northern Hemisphere, the southward flow along the bottom of the eastern margin of the ocean has less than the average salinity of the ocean. This is because of the fresh water input from rivers and, especially in the Atlantic, because part of the cold bottom water was originally melting ice water from the Arctic Ocean. Thus in the Atlantic, there is a seasonal variation in the cold bottom flow, more in the summer when arctic ice is melting and less in the winter. This seasonal variation will be increased ADV. The flow will be stronger when the earth is nearer the sun and slower when it is farther away.

Pacific Ocean circulation BDV is mainly driven by the westbound winds of the tropics. That is, in the Pacific, the BDV ocean flow more closely resembles a slow version of the atmospheric circulation and little mixing of the waters of the North and South Pacific occurs. The winds that drive the Pacific circulation tend to pile up water on the west side of the tropical Pacific in years when they are stronger than average. At irregular 3 to 5 year intervals these winds weaken and the weight of this accumulated water presses down on the deeper water causing it to move eastward along the bottom. When the eastbound deep water approaches the West Coast of South America, especially the westward continental bulge of Ecuador and Peru, it is forced to the surface. (It is replacing the surface water that the winds have transported towards the Philippines as well as bringing the ocean into more hydrostatic equilibrium.) This deep water is rich in nutrient and supports the fishing fleets of Peru and Ecuador, but it also cause droughts in some areas and excessive rain in others. The westward winds only slacken; they do not stop, so surface water continues to be transported to the west, but more slowly in reduced in quantity. Because this deep water rises to the surface and lingers longer when the winds are weak, the surface water becomes warmer off the West Coast of South America than normal. (Some textbooks erroneously imply that this warm surface water is warm water from the Western Pacific returning thousands of miles against the wind!) This phenomenon is called El Nino.

The difference in the nature of the Atlantic and Pacific Ocean circulation patterns is due to three factors. By far the least important is the fact that the Pacific has no gift of salt on its eastern side to help hold the cold water flow down as it warms on its journey southward towards the equator. Much more important is the fact that the Bering Straight (gap between Eastern Russia and Alaska) is relatively narrow and limits the flow of cold but fresher Arctic Ocean water into the North Pacific. (The North Atlantic opening between Greenland and Norway for bottom arctic water outflow and Gulf Stream remnant flow into the Arctic ocean is nearly 100 times greater than the Bering Straight.) This lesser flow also has a much greater expanse for its journey towards the equator, so it is much slower drift towards the equator than the corresponding bottom flow in the North Atlantic. This warms it more before it approaches the equator and explains why it rises to the surface in the Pacific Ocean but much of the corresponding southbound bottom flow in the Atlantic continues into the Southern Hemisphere.

Finally there is the fact that the tropical Pacific is six or seven times wider than the gap between Africa and South America, so the trade winds have much more "reach" (effect) in the Pacific. Oceanographers consider the Pacific circulation to be "thermally driven" (You and I might say "wind driven," but the winds themselves are thermally driven.) Also the density difference between tropical and north Pacific waters is mainly a thermal difference, not so significantly due to differing salt contents as it is in the Atlantic. They say that the Atlantic circulation is "thermohaline driven" to recognize of the significant role differences in salt concentrations play with what I have called the "piston effect" in the North Atlantic.

The reason that Boston's winter days can be colder than in Oslo is due to the Coriolis force, which is responsible for the fact that the prevailing winds come from the west at these latitudes. Even thought Oslo is 2000 km closer to the

North Pole, the climatic redistribution of heat more than compensates for the latitude inequality represented in the table at the beginning of this chapter. The land surface cools much more quickly in winter than the sea for the simple reason that the cold surface land can not sink and mix with the much warmer land only a few feet below. There are other factors involved, such as the fact that the heat capacity of the sea is twice that of the land, but it is the ability of the sea to mix vertically that is dominate. (Wind mixing of the upper ocean layers also plays a role.) Thus the surface waters off the East Coast of Norway are much warmer than the frozen land of the Midwest at Boston's latitude. Often the prevailing winter winds blowing into Boston have come from even colder landmasses in Canada. Nature will continue to do a superb job of moving heat from the sweltering summer of the Southern Hemisphere to the North during its winter, ADV. Norway will be one of the first to suffer the results of her success.

For BDV Norway it is not only the fact that the wind blowing across the relatively warm surface water of the North Atlantic ocean is warmed by contact with the ocean that is important. Very important is the fact that the dry air that left Canada is now moist air (stored heat of evaporation) which produces snow as it climbs up the mountains Norway. The abundant snow that falls in Norway in winter is partially responsible for keeping the air temperature near freezing while thermometers in Boston are falling well below freezing every time a dry Canadian air mass moves in. I am stating this again because it is difficult to over emphasize the effect of ocean transport of heat and <u>moisture it supplies to the air</u> has upon the land. BDV this transport was a blessing for Norway, but you can get too much of a good thing. ADV this climatic transport will be much stronger and produce a new ice age for most of the Northern Hemisphere. It will begin in Norway as is discussed in the next chapter.

Chapter 6

▼

Climate, ADV

The BDV earth is critically balanced between ice ages and the interglacial periods like the current era. Most experts attribute the transitions between ice ages and interglacial periods to slight periodic changes in the polar axis direction and slight natural changes in the eccentricity of the earth's orbit which, like the ice ages, have a period of approximately 100,000 years. Both George and Jack disagree. They think it is slight variations in the solar surface temperature that cause ice ages to come and end. Prior ice ages have begun slowly but ended in less than 10,000 years. The earth's orbital changes are steady cyclic changes. This is one argument that George and Jack use to support their view. Another is the fact that ice ages occur in both hemispheres during the same periods but polar axis tilting would tend to warm one hemisphere and cool the other, if it can have any effect on the average annual temperature. For me, however, their strongest argument is the fact that clean ice and snow reflect at least 80% of the incident solar radiation. Slight changes in the orbit or tilt of the earth could, at most, cause slight changes in annual solar heating. The reduction of solar heat absorbed by the earth when ice reflection during an ice age is sending most of the solar heat back into space would be much larger. Once the earth is in an ice age, it could never come out of it because of a very minor increase in solar heating that might be caused by the small periodically reversing orbital conditions.

Jack believes that our sun is a slightly variable star which has a very small but relatively sudden brightening approximately every fifty to one hundred thousand years. Without some external cause like Jack's solar brightening it is hard to understand how an ice age could end[4]. When ice covers large parts of the earth, the earth should be stably trapped in a permanent ice age. Jack's view is that ice age conditions are the normal state of the BDV earth and we are periodically rescued from the ice sheets only by a relatively rapid (a few thousand years or less in duration) brightening of the sun. Many variable stars do tend to brighten suddenly compared to their waning periods. Perhaps most stars, including the sun, show some small periodic variations in their surface temperatures. Jack even has some ideas as to the mechanism of the oscillation. Roughly Jack's idea is as follows.

Imagine that the outer layers of the sun were contracting slightly. The earth would receive less energy if the sun were smaller with the same surface temperature. As the outer layers fell, they would be converting gravitational potential into kinetic energy and heat, but the heat would be produced at the bottom of the falling layers, where internal pressure stops the fall, and not compensate for the reduction in size. The earth would begin to accumulate ice, but the sun would soon find a new static level and stop contracting. In the region immediately below the deeper and now hotter zone, the flow of energy to the surface is by convection. The magnitude of this convective energy flow would decrease slightly with less thermal difference between the interior and the surface. (Convection depends upon this temperature difference.)

The energy production in the solar core would not change immediately. Because energy transport deep with in the sun is a very slow process, the deeper layers would also slowly become hotter as the energy escape rate in the convective zone is lower. The heat capacity of the sun is enormous, but eventually the core where nuclear fusion produces energy would become hotter and energy production would increase. It would also take a very long time for this greater rate of energy production to reach the bottom of the convective layer and then thousands of years later the surface layer whose contraction had started the process. When it did so, this pulse of increased temperature/pressure would reheat the surface layers and cause an expansion of the surface. The larger and hotter solar

---

[4] An alternative to Jack's explanations as to how the earth emerges from an ice age is sometimes called the "white earth theory." This theory requires that the entire earth be covered by ice and is thus difficult to reconcile with the slow evolution of life on earth. (Without any photosynthesis while earth is ice covered, what would organisms eat?) The basic idea of the white earth theory is that volcanoes would slowly increase the carbon dioxide content of the atmosphere, eventually allowing the greenhouse effect to melt the ice. (The normal removal of carbon dioxide by absorption in the oceans and by green plants would not occur while the entire earth is ice covered.)

surface would quickly melt the ice that had accumulated on earth during the ice age which was caused 50 to 100 thousand years earlier by the initial contraction.

The expanding surface would be converting thermal energy back into gravitational potential and soon cease expanding and cool. With the gravitational energy restored, and a cooler surface, the ice forming part of the cycle could begin again. (A new earth-cooling surface contraction with very much later heating of the solar core and the associated increase in energy production that is not evident at the surface until much later still.) This type of oscillation model would explain why ice ages terminate relatively quickly when the energy pulse first reaches the surface making it not only hotter but also with slightly more surface to radiate until the next contraction starts.

If ice ages cycles begin with a slight contraction of the sun, the earth might actually warm a little soon after the end of the surface contraction phase because the higher temperatures produced at the bottom of the falling layers as gravitational energy is converted into heat would also propagate up to the surface. This could more than compensate for a slightly smaller radiating surface for a few hundred years. This "gravitational surface heating" would not last as long or be as great as the fusion heat produced by the core that eventually terminates the ice age. It is only an "Indian Summer" before the next ice age begins.

That is, initially the gravitational energy would be converted to kinetic energy of the "falling" surface and the earth would cool until the surface stopped falling because the surface area is decreasing and kinetic energy is being produced, not thermal energy, until the fall stops. Thus at the beginning of an ice age, the earth's temperature would first drop slightly for a few hundred years and then increase again when the contraction stops and some of the thermal energy propagates back to the surface.

The earth was cooling a few hundred years ago and the sun does now appear to be radiating more than it did a few decades ago. According to NASA, solar radiation is now increasing each year by a factor of 1.00005, so perhaps the contraction phase in one of Jack's cycles ended about 50 or 100 years ago. If so, the sun would soon send <u>both</u> hemispheres of the earth into a new ice age, if the dark visitor were not coming. (That is, the DV may be "God sent" to keep at least half the earth habitable.) In any case the current rapid rate (compared to the rate of evolution of life on earth or ice age cycles) of increase in solar radiation can not have existed for very long so Jack's idea that the sun is slightly variable star is supported by NASA's measurements. He admits that the mechanism he is suggesting is only a qualitative idea, without quantitative support, at present.

The geological record also indicates that the earth is "overdue" for the next ice age. Jack's contraction phase may have started in the 16th or 17th century. During the 17th and 18th century the average temperature of the earth was approximately two Fahrenheit degrees cooler. This period is called the little ice age because glaciers were growing, not shrinking as they are today. Thus Jack's solar contraction phase may last about 350 years and terminated at least 50 years ago as the solar radiation measured by NASA is now rapidly increasing.

In any case, the dark visitor will produce an enormously greater increase in the earth's eccentricity compared with the small change that occurs with a natural period of 100,000 years. It is certain that much of the land in the Northern Hemisphere will be under very thick sheets of ice in about 100 years, even if fossil fuel combustion were greatly increased. In this regard the dark-visitor's ice age will differ from all prior ice ages in that it will have an extremely rapid onset and will never end because this change in the earth's orbit is permanent, not cyclic.

Because the more eccentric orbit will persist for millions of years, it is difficult to imagine even Jack's periodic brightening of the sun will be able to terminate it. ADV the earth will no longer be critically balanced on the edge of an ice age. It will be in a permanent one, at least in the Northern Hemisphere. The objective of this chapter is to explain how this climatic change will occur and why the Southern Hemisphere will escape the grip of the ice.

The analysis that supports these predictions is buried deep inside several computer programs. George provided me with an outline of some of the ideas contained within these computer programs. The explanations I present in this chapter are based on that outline. The predictions presented come from the computer analysis of the climatic changes that will follow from the dark visitor's change of the earth's orbital eccentricity. These programs have evolved over the years with many people making contributions. Perhaps no one completely understands them.

After the dark visitor has passed, the year will have 378 days and the earth will be nearer to the sun than the current average distance for 148 days. (It travels faster when near the sun so this is less than half the year.) During most of this period the Southern Hemisphere is pointed more towards the sun more than the Northern Hemisphere. (The earth is tilted almost 24 degrees relative to its orbit plane.) That is, the Southern Hemisphere is in summer and the Northern Hemisphere is in winter when the earth is near the sun. (Even BDV, it was true that the Northern Hemisphere was closer to the sun in winter, but only by an insignificant amount prior to the passage of the dark visitor.) It is the tilt of the planet relative to the orbit plane that causes the change of seasons, not the difference in the separation from the sun, when the orbit is approximately circular. The ellipse of the ADV orbit will still resemble a circle. I will begin the description of why the climatic changes will occur with the earth in that part of the ADV orbit, when it is nearer to the sun than it was BDV.

Near Perigee Effects:

When the earth is closer to the sun the thermal input to the earth will be greater. During the day of closest approach, (Perigee) the earth will receive 13% more heat from the sun than the average thermal input before the dark visitor disturbed the orbit. In the Southern Hemisphere this summer heat will evaporate more water from the oceans. Some of this evaporated water will condense on Antarctic ice and assist the direct solar melting of Antarctic ice during the local summer.

It will take about 100 years for the Antarctica to be nearly free of ice and covered by green grass in the summer time, if the ice melts in situ (without quickly slipping into the sea). If it were not for the fact that the winter night there will be more than six months long and still very cold, it would be a pleasant place to live. If, however, large quantities of land supported ice slip into the ocean during the first decade ADV, many important cities in coastal regions, including Boston and New York will flood. Even thought most of the Antarctic ice cap will definitely melt, it is not certain that coastal cities will flood. In fact, if most of the Antarctic ice slowly melts on land without catastrophically slipping into the sea, Boston and New York will eventually be far from the sea, but no one will be living there then to care. Why all this is true will be explained in the paragraphs that follow.

Although the sunlight and thermal energy incident upon the earth will be greatly increased, especially in the Southern Hemisphere, during this "closer to sun" part of the earth's new orbit, Southerners will not be slow cooked. The fact that the Southern Hemisphere is approximately 90% water and 10% land is the main reason. The heat that reaches the surface will mainly evaporate more water instead of raise the temperature. During the hottest portion of the ADV orbit, the resulting cloud coverage will be nearly global with only occasional breaks in the clouds at mid and high-latitudes in the Northern Hemisphere. In this respect the earth will somewhat resemble Venus.

Much of the incident solar radiation will be reflected back into space by this seasonal coverage of clouds and never reach the surface. The southern summer cloud cover will also prevent most of the heat energy radiated by the earth from escaping into space. Thus the surface will be slightly warmer ADV even though less sunlight is reaching the surface because of its "blanket of clouds." Torrential rains will fall every evening in the Southern Hemisphere. Because of the melting Antarctic ice, the salinity of the southern oceans will not significantly increase for many years ADV despite the contracting total volume of the ocean. Equatorial surface water will be warmer ADV and easily float on the colder bottom water while this warm surface water is flowing northward into the Northern Hemisphere as some of it did, but to a much less extent, BDV.

During this portion of the ADV orbit, it will be winter in the Northern Hemisphere, but the winters will be milder than they were BDV for two reasons. First and most obvious is the fact that the Northern Hemisphere is also closer to the sun and much of it is also covered with a warming blanket of clouds. The second reason is that thermal transport of heat from the Southern Hemisphere will make the winters mild throughout much of the Northern Hemisphere. In the North Atlantic, as discussed in the last chapter, this convective heat transport by the warm equatorial water and the associated atmospheric moisture was very significant even before the dark visitor came.

England is much closer to the Arctic Circle than Boston, but seldom as cold in winter as Boston is, because it is also warmed by the remnants of the Gulf Stream. The harbors on the West Coast of Norway in the part below the Arctic Circle remain free of ice during the winter for the same reason. (See prior chapter for more details.) Even though the Gulf Stream is warmer when it is passing off the coast of Boston than when its surface remnants finally reach England, the fact that the winds are predominantly from west to east gives globally transported heat to England but not to Boston. England is humid and wet with frequent winter rains or snows precisely because of this global transport of heat. The West Coast of Norway has great snow falls and many small glaciers for the same reason, but even in mid winter most of Southern Norway has winter temperatures which are often just below freezing. ADV this heat and moisture transport will be greater as is discussed in the next paragraphs and a major portion of it will come from the Southern Hemisphere instead of the tropic region of the Northern Hemisphere as it did BDV. This "transport generosity" of the south will be the curse of the north—this atmospheric moisture transported into the north's winter becomes the ice of the new ice age.

Because of the increased solar heating, the thermally driven global circulation in both the atmosphere and the oceans described in the prior chapter will be increased. Climate will work harder to share the excess heat of the Southern Hemisphere's summer with the Northern Hemisphere. The mixing of both the ocean waters and the atmospheric air masses will be larger, much larger, in the Atlantic Ocean. Some of the energy from the south will make hurricanes which will be larger, stronger and more frequent each fall.

An enhanced Pacific Ocean flow ADV is one of the climatic adjustment nature will make in an effort to share the available heat between the two hemispheres. Both the closeness to the sun and the inclination of the South Pole towards the sun contribute to provide increased heating in the Southern Pacific during the winter of the Northern Hemisphere. The Southern Pacific will share some of this heat with the North Pacific. The pattern of Pacific Ocean circulation will be significantly changed with more trans-equatorial flow of warm water and atmospheric moisture. The northbound Kuroshio Current off the coast of Japan is the counter part of the Gulf Stream. BDV it was significantly weaker than the Gulf Stream, but ADV it will be stronger than the BDV Gulf Stream and part of the flow will then originate in the South Pacific. That is, ADV, the enhanced Pacific Ocean circulation pattern will resemble the BDV Atlantic with more mixing between the two hemispheres.

The Atlantic Ocean flow will dramatically change with very extensive cross-equatorial circulation in a giant figure eight pattern. (Cold water crossing deep under the warm surface flow near the equator.) Warm water from the Southern Hemisphere will flow from the West Coast of Africa across the equator on the surface towards the Caribbean, but turn northward because of the Corolis force to form a "Super Gulf Stream." This transported heat will melt Arctic ice and greatly increase the precipitation, especially in Western Europe. The melting arctic ice combined with the greater "piston effect" (See prior chapter.) as the edges of the Super Gulf Stream begin to sink off the east coast of Newfoundland, will greatly enhance the cold bottom current (no long a slow drift) toward the equator. As this cold flow approaches the equator, it will turn more to the west under the influence of the Corolis force than it did BDV. It will pass under the crossing surface flow of warm water emerging from the warm South Atlantic. The cold flow will continue southward along the East Coast of South America, slowly turning turn towards the southern part of Africa under the pressure of the Corolis force. East of the coast of Brazil, it will rise to the surface (because it has less salt and is no longer so cold) and then begin to rapidly warm.

The gift of salt it received from the Mediterranean outflow is not sufficient to keep it on the bottom as it warms. (Eventually there will not be any gift of salt for reasons described later in this chapter.) By the time it reaches the middle of the South Atlantic Ocean, it will be well mixed with the saltier surface waters. It will slow down as it spread out on the surface. This will permit it even more rapid heating in the summer of the Southern Hemisphere. As it approach the West Coast of central Africa, it will continue turning left and begin the giant figure eight course again. The heat transported north of the tropics and subsequent moisture transferred to the atmosphere in the Northern Hemisphere by the Super Gulf Stream will be at least six times greater than the BDV Gulf Stream. Most computer model outputs (different inputs) show it as nine or more times greater.

Even in the BDV orbit, the great snowfalls always came late in winter or early spring rather than the bitterly cold part of winter when there was little moisture in the air. With the extensive cloud cover due to the increased evaporation of the oceans, the moisture content of the air in the Northern Hemisphere will support great snowfalls all winter long. At the latitude of Washington DC, great quantities[5] of snow will fall in these mild winters from the clouds that often fill the sky.

North of the Arctic Circle the winter days will still be short or dark all 24 hours. It will still be very cold even when the earth is nearer to the sun. Recall the table presented at the start of the last chapter. The protective cloud blanket will not extend into the Arctic, but there will be dense surface fogs between Iceland and England all winter long. The central Arctic air mass will, however, remain relative dry and cold. That is, it will still be relative distinct air mass with little mixing, however, the greatly enhanced Super Gulf Stream flow will slowly melt the Arctic Ocean ice. The cold dry Arctic air mass will not provide enough snowfall on the ice covering the Arctic Ocean for this ice to long resist erosion by the remnants of Super Gulf Stream. If Henry Hudson were alive and searching for the Northwest Passage thirty or forty years ADV, he would be able to find it. If they have boats and store fish for the winter, Eskimos may still be able to live in the Arctic, enjoying milder winters than they had to endure BDV. They make a friend of snow, but may need to dig deep shafts into it if they want to fish in ice covered bays during winter.

Norway will resemble glacier bound Greenland in less than two decades ADV. By the time the rest of the Northern Hemisphere has a thick ice layer, Norway's fiords will be so thick with ice that they will be hardly distinguishable from the surrounding mountains. These beautiful fiords exist today because the land adjoining them is still rebounding upward from the weight of ice that held them down during the last ice age. Relieved of their icy burden, these mountains are still rising faster than the sea. A minor misfortune of the dark visitor passage is that in a few hundred thousand years, the western mountains of Norway and their fiords will once again be flatted down by the weight of accumulated ice, but not below the new sea level. The sea will fall faster than the mountains for several hundred years, but the fiords being full of ice, will not be visible. As this new ice age will probably never end, these mountains will not rise again.

Ironically the southeastern part of Greenland, which is now buried under a glacier nearly one thousand feet thick will slowly melt from contact with the ocean fogs by condensing the associated water vapor in the air. (Today, the ice in

---

5   The total winter snowfall for Washington DC during the first full winter ADV predicted by the computer model with the "most likely" values for the input parameters is 77 feet of wet snow! After that it gets worse for many decades, but becomes stable at about 50 feet per year once the ocean surface is reduced by its retreat from the continental shelf.

some parts of Greenland is still more than one mile thick—such thickness were common at high latitudes in prior ice ages.) Eventually this southern coastal part of *Green*land will be green in summer. Most of Southern *Ice*land will also be free of ice. It is too close to the path of the Super Gulf Stream to remain ice bound even in winter, but the northern shore will still ice over in winter. In year two hundred ADV (even with the most optimistic assumptions made in the computer model) all other land masses away from the new coastline and north of latitude 60° N. will be covered by an ice sheet about a thousand feet thick! In approximately 50 years, the ice over Washington, D.C., Madrid, Rome, Beijing/Peking, etc. (all other inland points near or north of 36° N.) should be at least several hundred feet thick. Washington, the state, will share Norway's fate as will Alaska and all of western Canada. That is, even though it will take a little longer, they will be covered with ice more than a thousand feet thick. The strengthened Kuroshio Current may keep Tokyo free of ice for approximately 60 years. Some warm water will flow[6] in the Sea of Japan, which is west of this island nation, and this may give Tokyo a decade or two more than New York before it is uninhabitable.—The Super Gulf Stream will only briefly postpone the icing over of New York City. NYC will be buried under ice as it is west of this heat source and rapidly becoming inland city, far from the sea. Some of the "sky scrapers" of NYC will be pushed over by ice flow in only two or three decades, and then completely buried in a few years. The Mediterranean will again become an inland lake, but it will not dry up leaving salt deposits as it has in prior ice ages, because moist winds from the west will increase the annual quantity of snow and rain falling in it significantly. In any case, it will no long make a gift of salt to the southbound bottom flow through the ice covered "Valley of Gibraltar."

Perigee Summary:

The coming ice age differs from all the prior ones in many respects. The most important one is the increased transport of heat and moisture from the Southern Hemisphere to the Northern Hemisphere during winter. Regardless of whether prior ice age were caused by slight changes in the earth's axis of rotation or changes in the solar flux, the two hemisphere were subject to essentially equal treatment and iced over together during all prior ice ages. The coming ice age is

---

[6] George no longer has convenient access to details about the depth of the Sea of Japan, so he was uncertain about the magnitude of this flow as the ocean is retreating from the continental shelf.

caused by a significant change in the earth's orbital eccentricity. (From 0.0171 to 0.0836) The Northern Hemisphere will be closer to the sun in winter and the Southern Hemisphere in summer. The heavy snows that fall during the North's mild winters will not fully melt during the following cold summer, when the earth is farther from the sun. It is this failure to fully melt in summer that causes the ice to accumulate annually until the oceans are smaller—levels below the continental shelf.

The summer of the Southern Hemisphere will be both hot and very humid. Most of the glaciers that now exist in the Southern Hemisphere will not survive. Only a few at the highest altitudes in the extreme south may survive. Great sheets of extremely thick ice will accumulate <u>only</u> in the Northern Hemisphere. As the ice cover grows, it will spread southward more rapidly because much of the weak summer sun's radiation will be reflected back into space by the continental ice sheet growing southward more each year. (Ice reflects about 85±5% of the sunlight incident upon it.) That is, the cold northern summers will become colder and melt even less ice with each passing year.

Eventually all of the ice now in Antarctic and at least the top two hundred feet of the oceans will be stored on the land as ice in the Northern Hemisphere[7]. In some locations, parts of north-western Europe and some parts of western Canada and adjoining parts of Alaska, this ice will become more than one mile thick! In California the ice will be less thick, perhaps only a few thousand feet where San Francisco was, but the weight of the ice will probably trigger earthquakes in the San Andreas Fault system. It is possible that the dark visitor will indirectly level San Francisco before it buries it under ice. Exactly how far south the ice will eventually go is uncertain. At least the southern part of "Expanded Florida" should remain ice-free and this new part of Florida will be larger than the entire state is now because of the retreat of the ocean from the continental shelf. All of Mexico and the coastal area near Brownsville Texas should also remain free of ice but the "port" of Brownsville will be useless—far inland from the much-reduced Gulf of Mexico. I doubt if even Texans will care, but Texas will also become bigger as the gulf waters move away from the current coastline.

---

[7] One final point of no real significance should be mentioned to correct a slight misstatement of fact that is presented in Jack's section. The day probably will not remain with precisely 24 hours. When Jack review my final draft, and read this section, he realized that this redistribution of water mass would slightly change the earth's rotational period and asked me to note this somewhere in the book. Jack has always been obsessed with being accurate, even when it is totally unimportant.

Near Apogee Effects:

If the earth were to remain in that part of the elliptical ADV orbit more distant from the sun than the BDV orbit there would be relatively little serious impact on the earth. There would be only be small changes in the ocean and atmospheric circulation patterns from those described in the last chapter if the earth were still in a nearly circular orbit approximately 5% farther from the sun than its current average separation. With the new orbit, it will never be more than 11% farther, so 5% is a representative value for the average of this more distant part of the new orbit. Obviously the earth would become colder because an earth farther from the sun would receive less heat. The year would also be longer. Even if the earth's orbit were circular at the extreme separation (apogee) of the ADV orbit, the earth would still be much closer to the sun than Mars is even when Mars is at its closest point of approach. (Mars' current eccentricity is 0.0934, but it too will change.)

Mars is very cold, but living on Mars would not be impossible if it had sufficient air to breathe and water to drink. Man's first settlements on Mars will no doubt be sealed enclosures and underground to a large extent. Recycling of water and nuclear energy would be essential to sustained life on Mars, but the earth is our current concern, not Mars. (Evacuating the Northern Hemisphere to Mars is neither attractive nor possible.) The possibility of life on Mars is mentioned only to reassure the reader, frighten by the conclusion of the prior subsection, that the dark visitor will not abolish all life on earth. The ADV earth will still be naturally habitable, even in southern parts of the Northern Hemisphere, and most of the Southern Hemisphere, possibly even including parts of Antarctica near the new coastline. That is, in parts of Antarctica's continental shelf that are now under water. The dry land produced in the Southern Hemisphere from the now flooded continental shelf will make a very significant increase in the habitable land of that hemisphere and should have unusually fertile soil. (The rains will quickly leach out the accumulated salt.) Continental shelf oil will be much cheaper. Even the blackest cloud (or the darkest visitor) can have a silver lining.

In addition to the fact that earth will still be much closer to the sun and still have a much thicker atmosphere, which makes it warmer also by the greenhouse effect, there is another important difference between earth and Mars. Three fourths of the earth's surface is covered with water. As discussed in the last chapter and in the prior section, the surface water will sink as it becomes colder and mix with the deeper water cooling deeper parts of the ocean and there is a tremendous quantity of heat stored in the ocean so the temperature drop will be small during the period the earth is further from the sun. A simple, but reasonably accurate, estimate for the average ocean temperature to drop on "apogee day," the day when the earth is farthest from the sun, can be made by answering two questions.

1) What is the solar heat reduction due to the fact the earth is father from the sun?
2) How much does the temperature of well-mixed oceans need to drop to supply the "missing" heat?

I am only trying to give approximate answers to a very complex question. George's more accurate models predict a slightly greater decrease because they include the effect of ocean circulation such as the Super Gulf Stream and treat the vertical mixing in the ocean more accurately than I can. I am also ignoring the effect of changes in cloud coverage, but this is not as bad an error as it may initially seem to be because while clouds reflect solar energy back into space, they also reflect radiant heat escaping from the earth back to the earth. The net effect is that a complete coverage by clouds more than compensates for loss of incident sunlight reflected back into space. The high surface temperature of Venus is proof of this.—The surface of Venus gets less sunlight and direct solar heat than earth's surface, yet lead will melt there! I.e. there may be "lead lakes" on Venus even though less sunlight reaches the surface there than reaches the earth's surface.

We can answer question one as follows.

1) Each square meter of solar radiation captured by the BDV earth supplies approximately 1000 joules of energy per second. There are 24x60x60 = 86,400 seconds in a day. The diameter of the earth is 12,750 km. Thus the cross section of the earth is approximately $\P(12,750 / 2)^2$ square kilometers or 127×10^{12} square meters. Hence the BDV earth gets approximately 10^{22} joules per day from the sun. It radiates away this same amount. (Actually very slightly more because of fossil fuel combustion and natural radioactivity.) If it were cooler it would radiate less, but as we are only trying to provide an estimate of the temperature drop we will assume that the full reduction in solar heat absorbed the ADV earth must be replaced by cooling the oceans. We have also neglected the increased cloud coverage that will help keep the earth from cooling but the fact that eventually the oceans will be smaller effectively cancels these positive factors. (I say that because this crude model is only slightly more optimistic than George's detailed computer analysis.)

The solar radiation intensity decreases as the square of the distance from the sun. Thus on apogee day, when the earth is 11% more distant, the ADV earth will receive 19% less from the sun. Hence the earth must cool enough to supply approximately 1.9×10^{21} joules during apogee day.

Now for the second question.

2) The BDV oceans contain 1.4×10^{12} cubic kilometers of water. Each cubic kilometer contains 10^{12} liters. When the heat stored in the land, atmosphere, rivers and lakes is considered, it is as if there were approximately 15×10^{23} liters of water available from which to extract heat, however, much of the deeper ocean is already too cold to be a source of heat if the cooling required to supply the "missing" heat is slight. I will assume that only 1% of the ocean and equivalent land, atmosphere and inland water is warm enough to supply the missing heat. (1.5×10^{22} liters) Thus there are approximately 7.9 liters available for every joule of heat "missing" during apogee day. Each calorie is 4.185 Joules and each degree centigrade that a liter drops supplies 1000 calories or 4185 joules per liter degree drop. From this crude calculation, we understand that even during the worst day of the solar deficit, the earth will cool only about $1/(4185 \times 7.9) = 0.000,03$ degrees centigrade. The total cooling during the 230 days the earth is farther from the sun than it was BDV would be less than 0.002 degrees centigrade. (about 0.003 degrees Fahrenheit.) Thus there is no danger that we will freeze during the first year while the ADV earth is farther from the sun.

The heat deficit incurred during this period will be partially repaid during the period when the ADV earth is closer to the sun than the BDV earth. Eventually, in a few centuries, the temperature will stop dropping with the annual average temperature less than one degree centigrade lower than it is today. Because of the complex effects of the greater cloud coverage and the high albedo of the ice-covered North, the annual change in the earth's heat balance is too small even for George's sophisticated computer programs to predict with confidence. His models predict it will eventually be approximately one-degree centigrade cooler on average, or a little less, depending on the set of assumptions assumed in the models. We are indeed fortunate that the earth is a "water planet" with a thick heat-retaining atmosphere.

Apogee Summary:

The reason that the earth will enter a new and permanent ice age for the Northern Hemisphere has very little to do with the slight negative change in the annual heat balance of the earth. (These calculations were the reason that Jack and George began to doubt the generally accepted cause of prior ice ages. The convincing arguments presented at the start of this chapter came later. As a historian, I like to keep things in their true sequence, even if I am not concerned with the precise dates.) At worst this heat balance factor alone (no change in circulation patterns) would be like a return to the conditions of the 17 and 18th century when the average temperature of the earth was also about one degree Celsius less. Glaciers were slowly growing then but they would never have claimed most of the landmass in the Northern Hemisphere as they will in the dark visitor's ice age.

The changing patterns of ocean and atmospheric circulation and the associated redistribution of water to the Northern Hemisphere during the portion of the ADV orbit discussed in the prior subsection are the primary causes of the new ice age. The contribution of this more distant part of the orbit to the Northern Hemisphere's disaster is simple. There is too little solar heating during the Northern Hemisphere's summer to melt all the snow and ice that formed there during the prior winter.

Inland regions:

Inland regions of the southern continents (Australia and the southern parts of South America and Africa) will not fully enjoy the moderating effect of the ocean's stored heat during their winters when the earth is farther from the sun. During their winters, while the Northern Hemisphere is experiencing cold summers due to the increased separation form the sun, these inland regions may cool up to 10°C (18°F) more than normal in their winter. (The computer models use Centigrade but I give the Fahrenheit equivalent for the American reader's convenience.) Unlike the ocean, the colder surface land can not sink and mix with the deep layers to remain relatively warm while the earth is far from the sun. Late winter temperatures in central and western Argentina and Brazil may drop to -15°C (-27°F) at night, and even lower in the high Andes. The summer cloud blanket that protected the Southern Hemisphere dwellers from a slow roasting and skin cancer will be absent during the winter. Hence the earth will lose heat to space more rapidly in the cold cloudless nights. Far from the ocean in the Southern Hemisphere, winter will be much colder ADV as the temperatures just given illustrate.

At Jack's observatory altitude, the winter will be much colder than the sea-level coast. It will resemble that of BDV winter in Minnesota. Ironically the ADV Minnesota winter will be warmer than it was BDV, but ice, more than a mile thick, will cover Minnesota with in about 300 years. (It takes a little longer, far from the moisture of the Pacific, for ice to get this deep.) Jack says he will dress warm and that "at least the air will be dry and free of clouds most nights." Prior to the passage of the dark visitor lower altitudes of these regions rarely experienced freezing conditions. Jack's new observatory will be useless all summer long. Karen says he can work with her in the rice paddy. I don't think she is kidding Jack! (A little real work will help you shed those extra pounds.)

Not only in the high Andes, but in most of the Southern Hemisphere when the earth is farther from the sun there will be few clouds during their winter. The Southern Hemisphere will experience much greater extremes between winter and summer after the dark visitor has passed, but life will still be possible almost everywhere there. (The Southern Hemisphere's summers will average about 2°C hotter and winters almost 10°C colder with greater variation far from the ocean in winter.) The increased summer temperature combined with the high humidity will no doubt kill thousands of old people with weak circulation and respiratory systems during summers that are much hotter than the average, but most people will be able to adapt. Many will take even longer siestas and begin working late in the night after the daily rains have finally stopped.

General Summary:

The most serious problem the dark visitor will cause is that the enormous accumulations of snow in the Northern Hemisphere's winter will not entirely melt in the following cold summer. When it is summer in the Northern Hemisphere the sun is farther away. Much of the Northern Hemisphere will initially be like Norway was before the dark visitor came. That is, glaciers will form first at higher latitudes especially Western Europe, Western Canada and Siberia) but as the years pass, they will grow, merge and expand southward. As they expand they will accelerate because most of the sunlight that does reach the surface through breaks in the clouds in the Northern Hemisphere's winters will be reflected back into space. That is, as the higher latitudes of the Northern Hemisphere ice over, the glaciers will grow faster and the summer will become colder, melting even less of the prior winter's ice.

In the initial winter of the Northern Hemisphere, just after the DV has passed through the solar system and the full effects upon the orbit have not yet occurred because the DV is still disturbing the orbit as it retreats into deep space, the milder winter (closer to the sun) and the significantly cooler summer (farther from the sun) that follows, will not seem to be as serious, despite the extremely deep snows, as the Southern Hemisphere's rapid transformation to more seasonally extreme weather with torrential summer rains and sever local flooding by rivers that will sweep whole cities away, but the Southern Hemisphere will escape the coming ice age. With in two decades, Boston will be covered with deep ice and uninhabited. Washington DC will also be in the grip of the continental glacier less than one decade later. With in two hundred years (or four hundred years with the most optimistic assumptions in the model) only the Gulf States, Southern Arizona, New Mexico and the extreme south of Southern California will be free of ice. (The San Diego area may survive the advance of the ice only to be destroyed by earthquakes induced by the weight of the accumulating ice.)

There is a great deal of uncertainty in some of these climatic calculations as the world has never been in such an orbit. Perhaps only the tip of Florida, the Key West Islands and Hawaii will be free of ice. The dark visitor may indirectly attack the first two of these "sparred" areas. The Super Gulf Stream's heat and moisture will spawn super hurricanes, perhaps at the rate of one per week, in the fall. Thus of all the US states, Hawaii may have the best chance of avoiding both ice and hurricanes; however, the more elliptical orbit will cause greater modulation of the gravity forces acting on the earth. This may trigger more volcanic activity in Hawaii.

Author's Disclosure:

Jack has reviewed my work and is pleased. He is giving me my choice of one of the three small farms he owns outside of São Paulo. I leave Boston in two days to make my selection. As selecting my farm will take some time, I may never return to the US. Thus I may be forced to recruit someone else to arrange for the publication of this book / report because I am told that getting a publisher can be a lengthy project. Good luck to all.—Billy T, 24 November 2003

Agent's Disclosure:

I received a letter from Billy T. It reports that Jack has observed the first occultation of a star by the dark visitor. When he observes another star disappear (briefly occulted by the dark visitor), he will be able to precisely calculate the speed and path of the dark visitor. (Note added to proof of this book.)—W. R. Powell, 11 January 2004

Part II

The DV Report

Jack wrote this report, but I have made minor editing modifications and some non-technical corrections, usually appearing in italics. Because I have already summarized some parts of his report for the non-technical reader and because his report became too disjointed when I tried to eliminate other redundancies, I decided to allow some redundancy in this book / report. Thus, Jack tells what will happen to the earth in his own words in the text that follows.

The following transitional fictional fable also was added. Except for name changes and some speculative invention in the first four chapters in order to make a coherent story with the known facts, this brief fable is the only pure fiction in this book.

A Clairvoyant Fable:

Although their leaders wore fine clothes, the common Crusader was dirty and dressed in animal skins. As they approached the city the Arab defenders thought they looked like bears walking beside the horseback men. The king asked his Persian poet / astronomer from where they came. He did not know, but he replied:

> Far north of the lion is the home of the bear.
> It is from there, near the tail of the dragon, that our danger comes.

He spoke more truth than he could possibly know.

Chapter 7

Discovery

Introduction:

This report is written for the scientifically literate to warn them of an impending event that will begin to significantly effect the earth in the second half of year 2008 (many very deep snows even that first winter.) It will drastically effect life on earth during the second half of 2009 and all the years that follow. An unknown object is approaching the sun. Most likely it is a small black hole, but other possibilities exist. Each will be discussed in the next chapter. The existence of this object can not be directly verified. It is not visible. Hence it is referred to as the Dark Visitor or DV. Now (Sept. 2003) it is far away, approximately eight times farther from the sun than Pluto. Its presence is known from the small gravitational effect it has had on the two outer most planets, Neptune and Pluto, especially Pluto. The next few chapters will describe what else is known about the DV, what is still uncertain, and how these facts were determined.

Perturbation & Discovery:

Small perturbation observed in planet Uranus's orbit permitted a remarkable accurate prediction of Neptune's location. This lead to the discovery of Neptune in 1846, only one degree from the predicted location. Uranus is less massive than Neptune and periodically Uranus is closer to Neptune than it is to the sun. Thus massive Neptune makes an easily measured (but still very small) disturbance to the orbit of Uranus each time the lighter Uranus overtakes Neptune. (Being closer to the sun, Uranus travels faster and "overtakes" the heaver Neptune.) The discovery of Neptune from observed perturbations of Uranus's orbit is the classical example of how calculations using carefully measured data about small perturbations of known orbits can lead to knowledge of a new object in the solar system. This is the method by which the DV was discovered, but its exact orbit (actually a passing trajectory) is still uncertain as it still can not be directly observed.

The current DV has had essentially no effect upon Uranus for three reasons. (1) Most important is the fact that the DV is still very distant from the sun; (2) The current location and motion of Uranus will cause Uranus to be on the other side of the sun from the trajectory of the DV when the DV is "near" the sun. (Near is placed in quotes because even when the DV is at its closest approach to the sun it will still be more distant than the orbit of Saturn.); (3) Uranus is more tightly bound to the sun than either of the two outer planets and thus is harder to disturb. These facts and the history of Neptune's discovery are mentioned to illustrate how successful prediction of an unknown body's orbit (or passing trajectory) can be made from the observation of perturbations of another planet's orbit.

The analysis to follow suggests that the DV has made a very small perturbation of the orbit of Neptune. This is possible despite the fact that the DV has not disturbed Uranus to any observable extent because. (1) The solar gravitational acceleration that keeps Neptune in its nearly circular orbit is only 41% as strong as on Uranus. Thus, Neptune is easier to disturb; (2) Neptune is closer to the trajectory of the DV than Uranus is. The DV's force on Neptune, although still very small compared to the solar force, is in such a direction that currently it would tend to pull Neptune farther from the sun and tilt its orbit plane northward. Eventually when it passes behind Neptune it will slow Neptune down and cause it to fall towards the sun in a more elliptical orbit, but it will never hit the earth.

When Voyager 2 flew by Neptune in 1989 many new facts were learned about Neptune and its moons, among them was the recognition that it was necessary to increase the previously accepted separation from the sun by 0.17%. This may not seem like much, but considering that Neptune has been observed for nearly 150 years, it is a substantial revision. A massive DV distantly leaving our solar system in the 1930s and 40s provides an explanation for how such a change in the orbital parameters could occur. No other completely satisfying explanation has been suggested. This earlier DV may also be responsible for Pluto's tilted orbit plane.

Very small perturbations in Neptune's orbit, observed in the second decade of the twentieth century, led to the chance discovery of Pluto in 1930. While the discovery of Pluto was a direct result of a year-long systematic search for an unknown planet in the area of the sky indicated by analysis of these perturbations, it is now clear that the mass of Pluto, even with its companion moon Charon, is much too small to account for the perturbations in Neptune's orbit that were observed. The motion that Pluto (with Charon) does produce in

Neptune is extremely slight, complex, and only is significant when they are close, an event which reoccurs only every 494.4 years.[8]

We now can prove that Pluto most certainly was <u>not</u> the cause of Neptune's perturbations. The next paragraphs show why and present one proof. The <u>discovery</u> of Pluto was caused by predictions of a new planet, that were based on perturbations observed in Neptune's orbit, but it was pure luck that Pluto happened to be approximately where they were looking. The fact that the <u>discovery</u> was caused by these predictions does not imply that Pluto was the cause of these perturbations. If an equally intense photograph search for an unknown planet beyond Neptune's orbit had concentrated on any randomly chosen section of the sky in or near the ecliptic plane, it is highly likely that some small orbiting object would have been found. Dozens of "planetesimals" are now known to a large extent by chance observations. If Pluto were recently discovered, by a chance observation rather than as the result of a systematic search for a predicted planet, it never would have been considered to be a planet. It would be just one more, larger than average, planetesimal. Pluto is in a highly elliptical in orbit, far from the plane of the ecliptic in which <u>all</u> true planets are found and very much smaller than any of the true planets. Pluto is smaller than the earth's moon. The mass of Neptune is 17.15 times <u>greater</u> than that of the <u>earth</u>. Pluto is never very close to Neptune. It was something else, not Pluto, that perturbed Neptune. The currently approaching dark visitor did not cause these early perturbations. (They probably were caused by an earlier, more massive one that passed far from the sun.)

The combined mass of the double planet,[9] Pluto / Charon or PC, is 435 times <u>less</u> than that of the earth! Thus the Neptune to PC mass ratio is 7460 to 1.

---

[8] This period can be most easily understood if one considers the Neptune/ Pluto system just after Neptune has passed Pluto. For it to pass Pluto again, it must gain 360 degrees on Pluto. The orbital periods of Neptune and Pluto are exactly in the ratio 2 : 3. (Their weak infrequent interaction is strong enough to preserve this ratio.) Thus 1.5 Neptune years after this "first passing" starting point, the slower moving Pluto has completed one orbit and is back to the starting point but the faster moving Neptune has completed 1.5 orbits and is exactly on the other side of the sun from Pluto. That is, it has gained exactly half of the required 360 degrees on Pluto. Obviously it will take another 1.5 Neptune years to gain the remaining 180 degrees, so they pass again after exactly 3 of Neptune's orbital periods or 3x164.8 earth years.

[9] PC is considered to be a double "planet" because the mass ratio of Pluto to Charon is only 6 : 1, a much smaller ratio than any true planet-to-moon ratio. The earth-to-moon ratio is 80 : 1 and this is the smallest ratio for any true planet / moon pair in the solar system.

Because Neptune's mass is 7,460 times greater than that of PC, it is hard for PC to disturb Neptune significantly, especially when one considers their great average separation. Neptune and PC are never closer than 17 times the earth-to-sun separation, a distance commonly called one Astronomical Unit, AU. By way of comparison, to prove that the idea that Pluto was the source of the perturbation observed in Neptune's orbit is completely ridiculous, I will now show that the earth "perturbs" the sun 6.5 times more than Pluto can perturb Neptune even when they are at their minimum separation.

The sun is 330,000 times more massive than the earth and their separation is always essentially 1AU. To compare the effect of PC upon Neptune with the effect of the earth upon the sun, in terms of their mass ratios alone, we must "move" PC to a 1AU separation from Neptune (like the earth—sun separation) and reduce its mass so as to leave the gravitational force of PC upon Neptune unchanged. Thus the "effective mass ratio at 1AU" would be 289 times greater than the actual mass ratio or equal to 2,156,012. This is 6.5 times greater than the sun / earth ratio and means that the earth perturbs the sun 6.5 times more than PC perturbs Neptune! The factor 289 comes from the fact that the force of gravity decreases as the square of the separation and the 17 AU minimum separation between Neptune and PC, (289 = 17x17). The 6.5 is the ratio of the mass ratios, (2,156,012 / 330,000 = 6.53). Thus, PC did not cause the observed perturbations of Neptune's orbit that led to the chance discovery of Pluto.

For most of Neptune's orbit the separation between Neptune and PC is much greater than 17AU. For example, one and a half Neptune years after Neptune is closest to Pluto, it is directly on the other side of the sun from PC. Then their separation is approximately 70AU and they have no effect upon each other. In contrast, the earth's mass is continuously acting on the sun from a distance of 1 AU. Thus the effective earth-to-sun mass ratio is usually very much more than 6.5 times larger than the PC-to-Neptune ratio. Both the mass ratio and the separation factors make the earth move the sun much more than PC can move Neptune.

An alternative form of the comparison in terms of a single relationship would be to conceptually move the earth to 17 AU from the sun and increase its mass 289 times to keep the earth's gravitational field at the sun constant so it would still apply the same attractive force to the sun. Thus the effective mass ratio for the sun / earth system if the earth were 17 AU away would be 1142 because with the effective mass of the earth larger, the sun to earth mass ratio would be smaller. (330,000 / 289 = 1142). Hence if the earth were at Neptune's distance from the sun, yet having the same force upon the sun as it currently has, then the sun would be only 1,142 times more massive that the earth. The sun would be more easily moved by the earth than Neptune is moved by the PC. Again we find that the earth would move the sun 6.5 times (7460/1142 = 6.53) more than PC can move Neptune.[10]

To restate the extremely important point demonstrated above: <u>The observed perturbation of Neptune can not be explained as due to distant and small PC mass alone</u>. The preceding paragraphs give <u>irrefutable quantitative reasons</u> why PC acting alone can not account for the continuing perturbations that have been observed in Neptune's orbit. The combined effect of PC plus that of one or more distant but massive DVs can explain the perturbations that were observed in Neptune's orbit beginning in circa 1920. It is a gross error to attribute these perturbations to PC alone. Pluto was found by luck, just as the other planetesimals have been. Just because an object exists where one was expected from perturbation observations does not imply that object is the cause of the perturbations!

Such perturbation logic and analysis is "backwards" and led to wide spread acceptance for many years of PC having much larger mass than it is now known to have. The orbit <u>and mass</u> suggested in this report for the currently approaching

---

[10] Actually it is the relative forces that are compared, not the relative motions. It is impossible to compare the relative motions in any simple way because they are of a very different nature, much more different than apples and oranges. The sun's motion, if caused by the earth alone, would be uniform motion in a circular path as earth and sun revolve about their common center of mass, which is only 1/330,000 AU from the center of the sun. The sun's diameter is 1/110 AU, three thousand times larger, so this center of mass is deep inside the sun, very near the center of the sun. As discussed in the prior footnote, PC disturbs Neptune's motion only very slightly once every 494.4 years!

The small movement of the sun by the earth is unobservable while an even smaller movement of Neptune can be observed, because: (1) The sun is not exactly round and has large surface flares that make the location of the exact center difficult to measure; (2) Jupiter's larger movement of the sun masks the earth's movement of the sun; (3) Viewed from earth with the aid of a telescope, Neptune is a small dim "star" (magnitude 7.8) with a precisely measurable location as it is essentially a point of light. (Neptune is too dim to be seen without the aid of a telescope—the unaided human eye can see stars with magnitude 6 or less.)

It is pure luck that in 1929 Neptune had recently passed Pluto when the search for a planet roughly behind Neptune began. If one were to now repeat a year-long search for a "planet" behind Neptune, with modern instruments, it is probable that many new planetesimals would be found, but we would not again make the mistake of calling them planets. There are thousands of these objects in the Ort cloud, more distant from the sun than Pluto.

DV is a direct response to the question: "What is required to explain all of the recently observed perturbations of PC?" To answer this question, an accurate computer model of the "three body" gravitational interaction was constructed. The sun, a planet in orbit and an object passing in an open trajectory are the three bodies. (*The model is discussed in Chapter 9.*) By trial and error, a mass, velocity and initial location were found for the DV that reproduced well the record of perturbations that have been observed in the orbit of PC (after Neptune's and the other planets' perturbations have been removed). Although this report is not conclusive proof that a DV with the particular characteristic suggested in it exists, it is at least an honest answer to a question that is usually ignored. Something like the DV reported herein must exist and is coming our way. It is not even the first to have passed by our solar system as the above arguments about the inability of PC to explain the perturbations of Neptune observed during the 1920s prove.

Small black holes, residual cores of many prior generations of now dead stars, may be more numerous than the current stars. Recent calculations show that there are currently more stars than grains of sand in all the beaches of the world! It would be surprising if small black holes never passed near our solar system, but still a rare event because the universe is so large. Perhaps the black hole now approaching (if that is what it is) was a companion to the one that passed in the 1920s. (Most stars do occur in pairs so their black hole residues should also.) More details about black holes and other possible dark visitor objects are given in the next chapter.

The "best fit" to the observational data on PC was then tested to see what effect it had on Neptune. That is, the planet in the three-body model became Neptune instead of a point with the mass and orbit of PC and the model was run again, first with the best fit DV and then also with zero mass for the DV as if it were not there. The difference in the results between these two runs of the model is the DV's perturbation to Neptune's orbit. By doing it this way, any small errors in the model tend to cancel out. Because the perturbation effect of the DV upon Neptune was never very large, it could not be reproduced exactly the observed perturbations, but it is approximately correct in magnitude and direction. The 1920s object would have had a slightly different trajectory than one derived from the currently approaching object, even if both were residues of the same pair of now dead stars. Part of the inability to reproduce accurately the disturbance to Neptune's orbit in the three-body model may also be due to the fact the perturbation that PC does make is not dynamically included. (In its current state of development, the model can include only three bodies. On cloudy nights I am expanding it for four, but do not wish to delay this report.)

Finally the model was run again with the earth as the planet and the best fit DV as the passing object to see the effect the DV will have upon the earth. These results are presented graphically in the last chapter of this report. (*Chapter 12.*) They can be summarized as follows. The length of the day will not change but the year will be 378 days long. Except for the legal complexities this will cause in contracts and the calendar etc. this, by itself, would not be a very significant change. The significant physical effect is that the orbit will be more eccentric. For slightly less than 230 days the earth will be farther than 1 AU from the sun and for a little more than 148 days it will be closer than 1 AU. At its most distant point the earth will be 1.11 AU from the sun and its closest point of approach is slightly less than 0.94 AU away from the sun. Prior to the DV, the maximum distance from the sun was approximately 1.03 AU and occurred during the summer of the Northern Hemisphere.

The climatic effect of this 8% increase in the maximum separation is not small. The new larger separation of the earth from the sun will still occur during the Northern Hemisphere's summer and this is the root of the most serious problem—a new ice age in the Northern Hemisphere. The snow that falls in the winter will not fully melt during the colder summers of the Northern Hemisphere; instead it will accumulate year after year in a self-accelerating process as it increases the albedo of the earth. The oceans will eventual retreat at least to the edge of the continental shelf, making the current ports of the world useless. This will cause hardships in the Southern Hemisphere as well. The main problem there will be the hotter summers will evaporate more ocean water and torrential rains will fall every summer evening, flooding many cities. This ocean evaporation is the source of water for the snow accumulation in the north. At least the top two hundred feet of the oceans will eventually end up as glaciers covering much of the land in the Northern Hemisphere, in many locations with ice more than one mile thick. This ice age will be worse than any of the many that have proceeded it as almost all of the ice accumulation will be confined to one hemisphere and because it will never end.

The "bottom line" is that the Northern Hemisphere will enter a severe new ice age because the mild winter's heavy snows will not entirely melt in the following summer. (Summer in the Northern Hemisphere will occur during the 230 days when the earth is farther from the sun and receiving less heat.) During the 148 days when the earth is closer to the sun, the Southern Hemisphere will have its summer. It will be hot and wet, but perpetual summer clouds will cover that entire hemisphere during summer and shield the residents from the stronger sun. The southern winters will be much colder than they currently are. They will be longer as the earth will be farther from the sun during the southern winter, but southern winters will be relative dry and the snows will not accumulate. The summer heat will melt most of the snow and the heavy summer rains will wash any remaining snow into the sea.

These effects in the Southern Hemisphere will occur more rapidly and there the new climate will stabilize there in a few years. Many people will die in the floods of the first summer after the DV has passed and the Northern Hemisphere will have seen only benefits in the first year (milder winters and cooler summers) but within a decade or two life in much of the Northern Hemisphere will be impossible. Details, including the mechanisms of how this will occur, should have already been presented in an earlier chapter on climate. (*Now in two chapters, 5 and 6.*)

The "orbit" (actually it is a passing trajectory), mass and speed characteristics of an "explanatory" DV must satisfy many constraints in addition to explaining the 0.17% unexplained perturbation to Neptune's orbit that had accumulated and was recently discovered by Voyager 2. An "explanatory" DV must also supply the "missing" part of the earlier perturbations in Neptune's orbit that lead to Pluto's chance discovery. ("Missing" due to the fact that the very small mass of PC and it great average separation from Neptune supplies only a small part of the forces required to account for the observations.) It must also explain the perturbations in the orbit of PC itself (actually the orbit of the center of mass of the PC pair). PC's orbit is being disturbed much more than can be explained by its currently decreasing interaction with Neptune. Its orbit has been "revised" at least six times in the 70 years it has been observed! Also an explanatory DV must be consistent with some things that have not happened. For example, as is discussed in the next paragraph, the separation between Pluto and Charon has not changed.

The need for these revisions in PC's orbit can not be explained by occasional collision of asteroids, as has been suggested, because the asteroid would hit only one (Pluto or Charon), not both equally, yet their separation has remained constant. While it is impossible to accurately measure their separation directly, one can be confident it has not changed because their rotational period has not changed. (The cube of their separation is directly related to the square of the rotational period. Thus if one changes, so does the other.) If perturbations in the orbit of PC were to be explained by occasional meteorite impacts, then their separation (and hence their mutual rotational period) would also have changed. Rotational periods can be measured very accurately if observed for a long time (many complete cycles). Charon and Pluto's orbit about their common center of mass has been observed for more than 20 years.

The PC mutual rotational period has been measured several times with high precision. It is 153 hours and 17 minutes. (This rapid revolution is because Charon is only 17.1 times Pluto's radius from Pluto. The moon is 60.3 times the earth's radius away and takes approximately 28 days to orbit.[11]) During twenty years of observation, approximately 1144 full PC revolutions about their common center of mass have occurred. If their orbital period were "in error" when measured twenty years ago by only one part in one hundred thousand, then their current orbit positions would be "wrong" by 0.01144 of a full orbit or more than

---

[11] The rotational period of the moon is slowly changing primarily because much of the earth's surface is covered with water. Some of the change is due to the fact that the moon is relative large (compared to all other planet/moon systems) and thus has some "tidal" effect even upon the "solid" earth. Despite the earth's distance from the moon, the moon is actually flexing the solid part of the earth but most of the dissipation of energy is in the ocean's tidal currents. As energy is being removed from the earth / moon system by tidal dissipation, the moon is slowing down and moving farther away. Long ago the moon was much closer to the earth and thus had a shorter period of revolution. When the moon was much closer, the tides were much larger, perhaps nearly a hundred feet of variation in the sea levels at the shore twice each day. The days were less than 24 hours then so high tide came more often. Clearly life forms that were well adapted to swimming were favored during the early history of the earth.

No one knows how long ago the moon was captured or expelled from the earth by a giant collision with a meteor or comet. (Some experts think the later possibility is more probable and that the Pacific Ocean basin is the "crater" that was created. The heat of the impact would have left the ejected rocks in a liquid state that coalesced to form the spherical moon we see today. Some other experts think the Pacific Ocean basin is the crater but that no impact occurred. Instead, a massive object passing too closely by the earth, tore the moon's mass from the earth less violently.—Perhaps this DV is not the first.) We do know, however, that even if the moon was captured (instead of formed from part of the earth, like Eve from Adam) it is a relatively recent addition to our sky. How we know this is discussed in another footnote of the next chapter that is concerned with "gravity gradients."

The next time you look at the moon, recognize that it has not always been there, is moving away, and no one knows how it got there, or when. All of the "explanatory" theories have flaws! As if to taunt us more, no one can explain why it looks so much larger when it is near the horizon than when it is high over head, but that is, an unsolved problem in psychology, not astronomy.

Charon's gravity at the surface of Pluto is too small to significantly flex the rocks of Pluto and there is no liquid water on extremely cold Pluto. Thus the "tidal" dissipation of energy in the Pluto / Charon system is proceeding very slowly. So slowly that meteorite impacts will be the dominant cause of any change in their mutual rotation period (and separation).

1%, which would easily be observable. In general, whenever many periods of a stable oscillator are observed, the period of that oscillator is very accurately determined. In the case of PC the period computed from the first 572 orbits can be compared with that computed from the most recent 572 orbits observed to see if any change has occurred.

Thus, another constraint upon the nature of the DV or what ever is perturbing the orbits of PC, is that it must still be far away, so that its gravitational field is pulling on both Pluto and Charon essentially equally. If the DV is far away and yet slightly effecting even massive Neptune, which is more tightly bound to the sun, then it must itself be very massive, comparable in mass with the sun, or larger. The DV obviously can't be a sun (star) as if it were and were close enough to explain the observed perturbations, it would be much brighter than the moon. Even the least educated person on earth would already know of its existence from his direct experiences, unless he were blind. The next chapter considers some possibilities for the nature of the DV that can satisfy all of these constraints.

Chapter 8

DV candidates

In this chapter several different classes of objects that can satisfy the constraint discussed in the prior chapter are presented in separate subsections. Each of these "class subsections" concludes with a summary of the arguments suggesting that the DV is a member of that class and of the arguments why it is not. Finally a WAG on the probability that it is a member of the subsection class is presented. Because some understanding of stellar evolution theory is essential to the discussion in several of the subsections that follow, this theory is briefly outlined before proceeding to the class subsections.

Stellar Evolution:

When a typical star, like our sun, has converted all of its core hydrogen to helium, the production of energy by nuclear fusion in the core stops and its core begins to contract. Some nuclear fusion continues in a series of thin shells surrounding the core, which are still rich in hydrogen, until they too have exhausted their supply of hydrogen. The surrounding shell of "burning" hydrogen combined with the conversion of gravitational potential into heat by the collapsing core makes the core hotter. As the successive shells burn hydrogen, more helium is produces and falls in upon the core compressing and heating it further. It was already much hotter than "white hot" (nearly 10^8 degrees K) but typically the core does not get hot enough to "burn" the helium core into other elements. To do this, a bigger star with more gravitational energy is required.

    Hot electrons limit the contraction of the core of a typical star to 1.4 solar masses or less. The details of this physic are complex but the basic idea is simple. As the densely packed electrons (which at these temperatures do not belong to any particular nucleus) bounce off each other they exchange energy and maintain a distribution of velocities that can be characterized by a single parameter, the temperature. The heavier nuclei do the same thing and these two temperatures are the same.

Temperature is the average energy of motion, but to have the same average energy as the nuclei the relative light electrons must move much faster. As the core contracts and heats, the pressure builds to resist further contraction, initially in direct proportion to the temperature (and density), but as the electron velocities approach the speed of light the pressure increases dramatically, preventing further contraction. This process and the initial mass of the star divide stellar evolution into three broad classes. Most stars keep these relativistic electrons in their core and evolve into "dwarf stars."

More massive stars develop such high pressures upon the core that the electrons are forced to combine with nuclear protons, converting them into neutrons. Neutrons have no electrical charge so they do not resist further contraction as effectively as the charged protons did. These neutron cores contract until the very strong but short-range nuclear forces prevent further contraction. Then the core is one gigantic mass of nuclear (not ordinary matter) density and stellar evolution has produced a "neutron star" which typically is also a "pulsar" for many years.

If the star was very big, it can produce a core so massive that even the short ranged nuclear forces are not strong enough to resist the gravitational burden of the weight of the outer layers pressing down upon the core. The collapse continues. When this collapsing core reaches half the radius it had at the time when gravity first began to overcome the nuclear forces, the outer layers of the core weight four times more. (Gravity is an inverse square law force.) So the collapse accelerates. There is no limit to this process. As the core shrinks, the battle between the nuclear forces and gravity tilts more and more in gravity's favor. It is hard to believe, but true, that just before the end of this process the entire star is smaller than the period which ends this sentence. No magician ever truly made a rabbit disappear, but nature is constantly making entire stars disappear. We call a mass collected into a point of zero size a "black hole."

Thus three evolutionary paths are possible for stars. The typical star dies as a dwarf star. A more massive star as neutron star. Very massive stars produce black holes. Some readers may be satisfied with this broad outline and may skip to the subsections that discuss the different classes that are candidates for the DV. Others, more technically inclined, may be interested in more details, especially about our sun, the fate of the earth and mankind, pulsars as little green men, and footnotes about color, the moon, and objects being ripped apart by "gravity gradients." All readers should at least read the footnotes, for some surprising facts.

Our sun obviously has less than 1.4 solar masses. Thus our sun will follow the "dwarf path" of evolution. As the core of a typical star contracts and heats, the radiation from it increases dramatically (as the fourth power of the temperature) but viewed from the outside, the star appears to be cooling down and expanding. I.e. it is entering the "red giant" phase. The extremely intense core radiation is entirely absorbed in layers too deep inside the star to be visible from the outside but it is the source of the heat and pressure that makes the outer layers of the star expand. Approximately 8 billion years from now, our sun will expand enough to "swallow" both Mercury and Venus. At the peak of its red-giant stage, it might even engulf the earth, but of course the earth would be entirely gaseous, including the iron core, not a solid ball if this happens. It is more likely that the expanding sun will not reach as far as the earth but "only" cover much of the sky with a hot "red"[12] disk. Even this lesser advance will strip the atmosphere away and boil the oceans into space so if man is to survive he must find a new home or build one he can move to keep a suitable distance from the sun as it evolves.

The hot outer layers of the sun that have expanded beyond the current orbit of Venus will continuing expanding into space and cool because the solar gravity is too weak at Venus's orbit to hold these hot gases. As deeper layers of the expanded sun are relieved of the weight of these leaving outer layers pressing down upon them, they too expand further and escape the gravity of the sun, but the sun does not disappear entirely into space. The shrinking sun is losing low-density gases and its core is still dense and hot. As the sun shrinks, the gravity at the surface is increasing. Eventually equilibrium is achieved between the tendency for hot gases to escape into space and the residual solar mass's gravity which pulls them back.

---

[12] If some intelligent creatures living near another star (perhaps distant descendents of humans) with eyes sensitive to same part of the electromagnetic spectrum we call visible radiation were to witness our sun in this "red giant" phase, they are not likely to call it (even in their language) the equivalent of "red giant" because the sun is not red, blue or any other color. Color is not a property of any object.—If you doubt this and continue to think the sky is blue, the sun is orange, apples are red, etc. then imagine that thousands of years ago a virus killed all people on earth except those whose DNA coded for daltonism (the most common form of color blindness—people that see objects most people call red as green). Then all the people living on earth today would agree that a star in this phase should be called a green giant because it is green in color, just as green as their hemoglobin blood, but for the sake of clarity, the common erroneous practice of assigning colors to objects will be continued.

Near the end of this process, the white-hot core radiation ceases to be absorbed in the outer layers that are drifting away. Thus for humans the color of the sun changes from red to white. (For bees and butterflies, whose eyes see ultra-violet light, it changes to ultraviolet color.) This small and much hotter sun is now in the "white-dwarf" stage. Actually for the more massive stars on the dwarf evolutionary path, it is a little more complicated. The shrinking core may gain enough heat from the conversion of gravitational energy into heat and from the continuing hydrogen fusion in thin shells surrounding the core to become hot enough (slightly more than 10^8 degrees K) to fuse the helium into carbon using unstable beryllium eight as a brief intermediary. They pass thought a second or even third red-giant stage before settling down as a white dwarf.

The final white-dwarf star stage has no source of energy and is losing energy with each photon that it radiates so it cools and slowly changes it color again passing through a red stage and then becoming a "brown dwarf" which finally cools so much that it no longer radiates electromagnetic energy that human eyes can see, but it will continue "forever" to radiate less and less each year. Eventually the peak of its radiation will be in the radio part of the electromagnetic spectrum, or even longer wavelengths if the universe last long enough. They are now "dead dwarfs" or for reasons given in the first subsection to follow, "gray dwarfs."

The sun is a typical star and thus there must be "Sagan numbers" of these dead dwarf stars drifting about in space, but not one has been observed, nor could it be, unless it happened to come close enough to our solar system to disturb the orbits of the outer planets. (The reader should not prematurely guess from this fact that the DV is a dead dwarf star.) While it is certainly possible that the DV is a modified dead dwarf, it is not the most likely possibility. The DV could be a very old core of a star in this "gray dwarf" stage if during the eons it has wandered in space, it managed to collect enough cosmic dust and other matter to nearly double its mass.

If a star has more than 5 solar masses, it releases enough gravitational potential energy during the core collapse to continue nuclear fusion beyond the hydrogen to helium stage, eventually converting most of the core in successive stage of nuclear "burning" all the way to iron. The initial stages of this process are complex but well understood thanks to Hans Bethe. These details need not concern us but explain why elements with even atomic numbers are much more common than those with odd atomic numbers and other details concerning the relative abundance of the elements less massive than iron. When most of the core is iron, energy production by nuclear fusion stops. (For an iron nucleus to fuse with anything else requires energy instead of releases energy—I.e. iron is the bottom of the nuclear well.)

The dynamics of these last stages of core collapse are not well understood and probably depend upon how spherical and homogeneous the core was. Because the rate of nuclear "burning" is strongly dependent upon the local temperature, any small region of the core that happens to be hotter than others will produce energy and convert matter more rapidly than its neighbors. This self amplifying instability has its limits via heat conduction and the fact that the hotter plasma becomes less dense and this lowers the burning rate but significant inhomogeneity with in the final "iron" core can be created by this and by convective mixing which at times can reach down to the core and may not do so equally on all sides.

Once the core is mainly iron it begins to collapse again and this time no new nuclear "burning" stage is ignited to stop the collapse. If this collapsing core has approximately four or more solar masses part of an inhomogeneous core may be thrown outward by the gravitational energy released by the collapse itself along with all of the remaining outer parts. (These ejected pieces of iron are probably the origin of some of the iron meteorites.) If the remaining core has approximately three solar masses or more, it will collapse to a black hole, a single point. Relatively homogeneous cores with slightly less than three solar masses may also produce these point-mass singularities while others with more than three times the solar mass may not because of their inhomogeneities etc. Unlike the relative sharp 1.4 solar core mass limit for star to go through the "giant-to-dwarf" evolution (which is set by the exclusion principle of quantum physic and the fact that even hot electrons can not move faster than the speed of light), the evolutionary path of more massive stars is less predictable.

The strength of gravity as a single point mass is approached increases without bound. At some spherical surface surrounding this point mass, the gravity is so strong that nothing, not even light, can escape. Anything that comes inside this sphere, including photons of light, will never leave. Thus this sphere has absolutely zero reflectivity. If you were to shine a flashlight beam on it only one second before you fell inside[13], you still would not see it because not one of the photons in your flash light beam would return to your eyes. This sphere is much

---

[13] This would not be possible, even if you went to the black hole in a steel spaceship. The gravity on the side of your body closer to the black-hole sphere would be so much stronger than on the more distant side that you would have been ripped apart hours earlier. It is this "gravity gradient" in a much weaker form that explains why there are two ocean tides each day, not just one. It is true, as most people believe, that the high tide on the side of the earth near the moon is caused by the moon's gravity pulling on the ocean water more than the average pull on the solid earth. The average

"blacker" than the blackest paint ever produced. This is why the sphere (not the point) surrounding the single point where the mass is concentrated is called a black hole.

Black holes are none the less more easily detected than the dead dwarf stars. Strong indirect evidence exists for several, but all are much more massive than dwarf stars. Black holes can merge to become bigger. That is, their "no escape sphere" is even bigger. All of the mass will eventually be in a single point of zero size, but this may take a long time. Typically the two black hole points will orbit each other to conserve their angular momentum, making bumps on the no escape "sphere" which will be radiating gravity waves and losing energy. (Either the gravity waves carry angular momentum or when only one point exists, it has some form of intrinsic angular momentum, like many fundamental particles do, called "spin," but this is beyond my field of knowledge and getting too speculative to continue in this factual report.)

Every galaxy probably has an enormous black hole with mass millions of times greater than the sun at its center and they are constantly growing bigger as they swallow whole solar systems that happen to come too close. Perhaps the most likely possibility for the DV is a small black hole formed from an inhomogeneous

---

gravity pulling on the solid earth is approximately equal to the strength of the moon's gravity at the center of the earth, which is an earth's radius farther from the moon and thus in weaker gravity than the ocean on the side near the moon. The ocean water on the other side of the earth has even less moon gravity acting on it than the gravity at the center of the earth. Thus the high tide on that side is not due to the water being pulled "up" by the moon. Instead it is due to the solid earth being pulled "down" or away from the water. It only appears that the water is rising because we assume that the earth is not moving in response to the moon's pulling. Both earth an moon are in fact moving about their common center of mass (as well as about the sun, and with the sun about our galactic center, and with our galactic drift in space, and with???, etc.) but if we neglect all these other motions, then after approximately 28 days, both moon and earth will be back in their original positions. The bulge of water we call high tide on the side of the earth far from the moon provides just enough hydrostatic force towards the moon to compensate for the weaker force of the moon. Acting together, these two forces keep the ocean on that side also in the same circular rotation path (about the common center of gravity of the earth moon system) that the solid earth is following. The fact that part of the earth is dry land makes things more complicated than explained here where tacitly a solid-core, non-spinning, earth 100% covered by a uniformly deep ocean has been assumed.

A footnote in the prior chapter explains why the moon is slowly moving away from the earth, so it is not precisely true that "approximately 28 days, both moon and

stellar core with approximately three stellar masses or a little more if it was very inhomogeneous. Like the dead-dwarf DV, the black-hole DV would be residue from an earlier generation of stars. Thus earlier stars with either more than approximately 4 or less than 1.4 solar masses may be the origin of the DV, but neither is a natural match to the mass of the DV. (The dwarfs are too small and the typical black hole is slightly too big.) After a brief "supernova detour", we will consider isolated stars with masses between these limits.

It is not important for our understanding of the DV, but it should at least be mentioned that enormous release of energy when the iron core of a massive star collapses produces the spectacular event known as a supernova and forms the elements heavier than iron. The suddenly released energy by gravitational collapse of the iron core drives a shock wave through stellar mass surrounding the core that is thousands of times stronger than any generated by high explosives. The entire non-core parts of the star and possibly some of the core, if it was inhomogeneous, are heated to extreme temperatures by this shock wave and hurled far from the imploding core. The expanding supernova star is so hot that most of its radiant energy is X-rays! Even in the visible light spectral range the large hot supernova can radiate more light that all the other stars in its galaxy for a few weeks.

---

earth will be back in their original positions." That footnote states that no one knows when the earth got its moon, but we can know it was not in the early history of the earth because of the same gravity-gradient phenomena discussed in this footnote. That is, if the moon is now moving away then 10,000 years ago it was closer and 10,000 years before that it was even closer as the tidal dissipation which is making the moon move away was greater when the moon was closer, but the moon was never "skimming" just above atmosphere as you might think if you calculate this moon motion backwards in time and ignore gravity gradients. Even if you assume that the moon is made of the strongest rocks known, when it gets too close to the earth the force of the earth's gravity on the half of the moon closer to the earth is so much greater than the force on the other half that the moon would be pulled apart by the gradient of the earth's gravity. Thus we know that the moon is not as old as the earth. We got it relatively recently, but exactly when is impossible to know and even the how it could have happened is difficult to explain. (Experts do not agree.)

Don't worry about being ripped apart by gravity gradient before falling into the black hole. It won't hurt. Your steel spaceship is much longer than you are thick, so the front end of it will be in much stronger gravity than the rear end as you approach the black hole. It will be ripped apart long before you are. Without any spaceship, you will already be dead before you are ripped apart. Fortunately this is just a thought experiment. None of it is real.

In year1054 a supernova occurred in the constellation Taurus. The remnants of its expanding gases are still hot enough to be visible today. They are called the Crab Nebula. At its peak this supernova star was so bright that it was visible during the day. In 1006 a supernova 100 times brighter occurred. It was a point light in the sky as luminous as the quarter moon, easily visible during the day and casting sharp edged shadows at night! If a comparable event were to occur in a close star, the X-ray flux and resulting ozone concentrations would probably make most life forms on earth extinct. Some soil bacteria on the shadow side of steep mountains, ozone resistant primitive plants and some deep dwelling sea creatures would probably survive to start the whole evolutionary process over again. It may have already happened to earth more than once before. If so, it could explain why we think life on earth started so late (only recently) in earth's long history. I.e. like our sun, current life forms may not be the first generation to follow an evolutionary process on earth.

These speculations about evolutionary history are well outside my field of knowledge and thus may be completely wrong. Religious people will certainly think so. What I am reporting about the DV is likely to be correct but some of the minor details are still uncertain. Religious people may think the DV is bringing the end of the world as many of the events predicted in the Bible to proceed this event appear to be happening, but I do not think so. Many people living in the Southern Hemisphere will survive. In this report, I am providing the reader with the arguments and associated quantitative facts (which he can easily check) that lead to the conclusion that the DV (or something very much like it) is coming. No qualified astronomer will deny that something other than PC is required to explain the perturbations observed in Neptune's orbit.—Read prior chapter, if you have not already done so, for one simple quantitative proof of this fact.

The stellar core left at the center of the Crab Nebula did not have sufficient mass to collapse to a single point even though calculations from the nebula mass seem to indicate that it could have. Perhaps at the start of the collapse, the core was very inhomogeneous and much of it was also blown away during the implosion. The residual Crab Nebula is not very symmetric. Perhaps the original Taurus star was not more than five solar masses as the calculations indicate. Instead of becoming a black hole, that Taurus star became a "neutron star."

If a star with five or more solar masses fails to produce a point mass singularity, at least part of the core is very likely to end up as a "neutron star." Cores with more than 1.4 but less than a little less than 3 solar masses usually end up as neutron stars. (They are too big to become dwarfs and too small to become black holes, unless members of a stellar pair.) In a stellar core that is becoming a neutron star the hot heavy nuclei are so closely packed together that the electrons which neutralize the positive charge of their protons do not "belong" exclusively to any nucleus. In some ill understood way most of these electrons combine with the nuclear protons to neutralize each other's charge resulting in one gigantic, extremely dense, neutron mass. As neutrons, the mass of our sun can be packed inside a ball less than 30 km in diameter. (That mass is now in a ball 1,400,000 km in diameter. The volume and density change is even more impressive as it is the cube of this ratio.)

Neutron stars usually are very rapidly rotating—most of the 1000 that are known complete many rotations each second but some take several seconds to rotate. A small, but rapidly spinning, body now carries the angular momentum that was in the much larger spinning core when it began to collapse. For quantum physics reasons that are difficult to understand (or not very clear?), this mass of neutrons is a super conductor or at least a very good conductor of electricity. This means that when the core collapses it retains and compresses the magnetic field it had when it was larger. This field is shaped like the earth's magnetic field, but tens of millions times stronger at the magnetic-pole surfaces. Like the earth, these magnet poles are not aligned with the axis of rotation. This intense magnetic field spinning around the axis of rotation in less than a second acts as an extremely powerful electric field generator, producing and accelerating electrons of both signs (positrons). These electrons are expelled from the polar region by both the electric field and the fact that magnetic pressure moves them to less concentrated field regions. Some recombine to produce half Mev gamma rays.

As the electrons leave they gyrate rapidly around the magnetic field lines. This produces an intense continuous beam of synchrotron radiation at the frequencies of gyration. This beam emerges from each magnet pole like the beam from a lighthouse. (It projects out along the polar magnetic field lines.) We receive only a very brief pulse of electromagnetic waves each time the continuous beam happens to sweeps across earth as if we were looking at a distant lighthouse. Thus the pulse rate equals the rotation rate of the neutron star. The pulse frequency (actually an energy distribution in a frequency band) is the electron gyration rate distribution / bands.

Most such narrow radiant beams never sweep across the earth. Instead they trace out circles in the heavens that do not include the earth's location, but we now have enough data to estimate that there are at least 100,000 of these "pulsars" in our galaxy. When these extremely regular radio pulses were first accidentally discovered (in 1967 by a female graduate student—I promised my friend, the politically active historian who is assembling this book, that I would emphasize this fact.) their source was unknown and some though they were signals from extraterrestrial intelligent beings. Consequently what we now call "pulsars" were initially called LGMs (for Little Green Men) but this name was dropped in a few months when a more systematic search revealed several more.

If the stellar core is in the 1.4 to 2 or 3 solar mass range, the process of electrons combining with protons to form neutrons and the gravitational collapse also releases enormous quantities of energy in an extremely short time, but far less energy than the supernova process a bigger star that has an iron core can undergo. When this star flares up many magnitude orders of brightness it is called a nova, not a supernova. Novas are called novas because usually, prior to their nova stage, unless they happen to be near us in the galaxy, they are too dim to be seen by the unaided human eye (I.e. not magnitude 6 or less). Thus when they enter the nova phase, a "new" star appears in the heavens, but it will fade away in a few weeks and only be visible with the aid of a telescope. Supernovas can be visible to the unaided eye for a few years. Now we must consider star pairs.

Almost all stars are found in compact groups and many stars are pairs in close proximity to each other. Then their mutual evolution can be much more complex. Often both formed at about the same time from the same cloud of hydrogen gas. Typically one will be little bigger than the other but both will resemble the sun in size, as much bigger stars are less common. Most of a larger scale density fluctuation in a gas cloud must end up in one star instead of several to make the typical big star. Small-scale density fluctuations are more common so "sun like" stars are much more common. (Astronomers first removed the earth from the center of the universe. Now we insist that there is nothing special about our sun! No wonder we are poorly paid, if paid at all.)

The bigger star of the pair will evolve more rapidly than the smaller one because its core is hotter having benefited from greater gravitational collapse energy. The hotter the core the faster the hydrogen is converted to helium. Thus it is not uncommon for the bigger member of the pair to already be in the white-dwarf stage when the smaller one is entering the red-giant phase. As the red giant's outer layers, which are still rich in hydrogen, expands towards the extremely hot surface of the white dwarf, and are attracted to it by its gravity, the white dwarf may collect enough hydrogen from the expanding outer layers of the red giant to come to nuclear life again. Isolated white dwarfs just slowly cool as they have no more capacity to generate energy, but a white dwarf favored by its red-giant companion with a new load of hydrogen is likely to become so excited by this gift that it explodes as a nova, not as supernova, even if initially it had too little mass to become an isolated nova.

DV as a FGD (Fat Gray Dwarf):

Although the DV is definitely not a star like our sun, it is possible that it is a type of star one stage older than the well-established classes of "brown dwarf" stars. Brown dwarfs should age into "gray dwarfs." They are given this name here for two reasons. (1) Gray is associated with very advanced age. (2) More seriously, because the radiation they emit is that of a gray body with temperature too low to emit visible light, or even detectable infrared radiation. Their radiation could be part of the cosmic background noise in the radio or microwave part of the electromagnetic spectrum. They have thus far escaped detection.

Gray-dwarfs stars were once like our sun but formed much earlier than our sun, probably in the first generation of stars after the "big bang." Our sun is at least a second-generation star because it contains small amounts of elements beyond iron in the periodic table that can only be formed in the violent death of prior generations of stars. Big stars age much more quickly and die more violently than small stars like our sun. Thus currently there are stars from different generations in existence, but most are relatively small like our sun, partly because they last longer and partly because they can form from smaller condensations of cosmic hydrogen which are statically more common than large-scale fluctuations in cosmic hydrogen clouds. Big stars can not calmly evolve into gray dwarfs, as our sun will if the universe last long enough. (Widely accepted current scientific theory agrees with the Bible that time has an end as well as a beginning.—See Stephen Hawking's *Universe in a Nut Shell* for more details.)

Although gray dwarfs have never been observed, the same standard theory of stellar evolution that is used to infer their existence also limits their mass to be less than 1.4 times that of the sun. The best fit to the observed planet orbit perturbation facts implies that the DV's mass is 2.2 times more massive than the sun. This is a serious reason for rejecting the idea that the DV is a member of the gray-dwarf class, but it is not the only objection (see next few paragraphs), however none of the objections permits us to eliminate this class of objects. The DV could be a very old and now "fat" gray dwarf if during the eons it has wandered in space, it managed to collect enough cosmic dust and other matter to nearly double its mass. Hence the FGD title to this subsection.

If the DV is an FGD, then it surely will become visible soon in reflected sunlight with moderately large telescopes. Probably it already is visible with the largest telescopes as anything made of ordinary matter with a mass of 2.2 times that of the sun should be visible even if it is currently 230AU distant from the sun. The quantitative arguments of the next few paragraphs yield this result. The reader not interested in mathematical details can skip the next two paragraphs, but should read this section's summary.

The FGD's average density would depend upon the heat still retained deep inside it and what material it collected to become "fat." The most common normal matter available is hydrogen. Probably a FGD would slowly collect hydrogen; slow enough to not become a star again. If we assume that it has the same average density as liquid water and require its mass to be 2.2 solar masses,

(4.4x10^33g) then its diameter is 2x10^6Km from fact that 4.4x10^33 = (4¶/3)r^3 where r is in cm. The sun's diameter is 1.4x10^6Km and at 1 AU this corresponds to half a degree. Thus the DV at 230AU should present an angular diameter of 0.0031 degrees. (0.5x{2/1.4}/230) This is essentially the same angular size as Neptune when viewed from the sun, however the sunlight reaching the DV would be less intense than that falling on Neptune by the ratio of (230/30) squared or 59 times weaker. To be visible, this sunlight must reflect and get back to earth—another intensity reduction of approximately 59. The intensity is also reduced by the reflectivity coefficient. A value of 0.35 for this coefficient is both typical and convenient because then the light coming back to earth is exactly 10 thousand times fainter than that from Neptune. (0.35 /{59x59} = 0.0001) I am also assuming that the refection is diffuse so the full disk reflects light back to us. If the surface is optically smooth, then very little light will return to earth. If it is approximate smooth, then even the biggest telescopes may not yet be able to see it. (If the moon were a spherical mirror it would be more like a star and provide much less light to earth.)

Neptune appears as a star of magnitude 7.8 and each increase in magnitude is a reduction in intensity by a factor of 2.512. Thus 10 magnitude steps are required to make a reduction of 10 thousand. (Exactly 10 magnitude steps because, by definition, five steps are a factor of 100. This is origin of the strange factor of 2.512 that is not exactly correct.) Hence with these assumptions the DV would now appear as a "star" of magnitude 17.8, too faint to be seen except by the larger telescopes. Perhaps in the entire world there exist 100 telescopes that could see this type of DV now. When the DV is only half as far away (2006) it will be 16 times brighter (magnitude 14.7) in reflected sunlight and many modest size telescopes should be able to see it, but by that time it will have already slightly lengthen the earth's year and made many large perturbations to the more distant planets.

It is highly unlikely that any telescope that could currently see a DV of the FGD class (with the above assumptions as to its size, reflectivity, locations etc.) is surveying broad regions of the sky looking for new objects. Almost all are employed at high magnification to look at a very small part of the sky that contains some very distant object of cosmological interest. Their light gathering power is required to see these distant objects. Because the light from these distant objects had to travel for so long to reach the earth, it left the source shortly after they were formed. Only the large telescopes are capable of generating optical information about the origins of the universe. No management committee of a large instrument is going to waste their unique ability by allocating telescope time to routine sky surveys that lessor instruments could do. Thus the fact that the DV has not yet been seen is not a strong argument against it being a member of the

FGD class. If it is not seen within a few years, however, this class can be definitely excluded and the black-hole class becomes more likely.

FGD Summary:

Current theory of stellar evolution sets an upper limit of 1.4 times the sun's mass on the cores of stars that evolve with age from "red giants" to "white dwarfs," then to "brown dwarfs" and finally to the "gray dwarfs" postulated here. It is for this reason that a gray dwarf does not seem to be a likely candidate for the DV that has approximately 2.2 solar masses, not because it has not yet been seen. Approximately 90% of all the mass in the universe is unknown "dark matter." Consequently these undetectable gray dwarfs may be extremely common, although "dark matter experts" do not often advance this possibility. The strongest argument in favor of the DV being a FGD is that there must be very many gray dwarfs and most have had a long time to collect cosmic dust. The probability that the DV is a FGD is set as 0.15, a WAG.

DV as a Neutron Star:

The DV could be a neutron star, but this is not very probable because all pulsars are neutron stars and the converse is also probably true. If it is true that all neutron stars are pulsars, then it is unlikely that the DV is a neutron star, but until we can be certain that the converse is true, this possibility can not be excluded. Even if the converse is true, the DV could be a very old neutron star. One fact about neutron stars that was not mentioned in the first section about stellar evolution is that their pulse rate is not strictly constant. As this star radiates a powerful beam, it must continually lose energy. Now that the pulse rates of many neutron stars have been observed for years not only is this fact confirmed by the steady increase of the interval between pulses, but also occasionally there is a sudden change in pulse rate corresponding to a slight increase in the spin rate. Perhaps the prolonged steady rotational slowing of the neutron star reduces the centrifugal force to such an extent that the surface collapses inward a little. Thus the spin rate would increase to preserve the total angular momentum during this subsidence of the surface.

The nuclear force, which is preventing more contraction, is not a "brick wall." That is, there would be some compressibility associated with a mass of neutrons. It should be mentioned other more complex explanations for the sudden "glitches" observed in pulse rate are being sought. It should also be mentioned that the electrons gyrating about the magnetic field lines are generating a beam, which is circularly polarized. This beam is constantly carrying angular momentum away and this is another reason why the neutron star's spin rate is slowing. Gravity won the battle and compressed the star to nuclear density. It certainly will not give up its victory to let the pulsar expand. If it can not expand, the fact that it is steadily slowing it spin between "glitches" represents a reduction in its angular momentum but the amount "missing" must be exactly the amount the beam steals away.—Conservation of angular moment is not violated. Thus even if all neutron stars are initially pulsars and thus easily detected when "only" 230 AU away, the DV could be a very old pulsar that is spinning too slow to now make propagating electromagnetic waves we would detect. Many pulsars have approximately 2.2 solar masses and this is an argument in their favor.

Neutron Star Summary:

The strongest argument against the DV being a neutron star is that no strong periodic pulses have been detected. No one would need to be specifically searching for these signals because at a range of only 230AU they would be so strong, even if "way off the beam center," that they would very likely be causing interference with some radios or radars etc. The first pulsar was accidentally discovered precisely because it was "interfering" with a radio astronomy telescope, but this discovery was the most important thing that radio telescope ever did and it is certainly astronomy, so it is a little strange to say that the first pulsar was "interfering"[14] with it. It was just an unexpected use of the new instrument. Astronomy is a long history of surprises as the 20% probably assigned to a DV of the "unknown-unknowns" class discussed in the last subsection of this chapter indicates.

The strongest argument for the DV being a neutron star is the typical neutron star has essentially the same mass as the DV. Also there is the possibility that some neutron stars may not be pulsars even initially. Perhaps some are not good conductors and thus do not trap and compress their magnetic fields (or never had one) as they collapse to nuclear density. Also there is the possibility that the beam has a more rapid intensity decrease with angle from its center than current theory would suggest. (The radiation emitted by the gyrating electrons is along the local magnetic field lines and even at the poles they are curved.) I.e. the beam could be

---

[14] "Interfering" is also not quite the correct term because the pulsar signals were not strong enough to block other signals as one normally intends when one says that something is "interfering with TV reception" etc. The pulsar signals were just "riding" on top of other signals in the waveform.

more like a laser beam than a car headlight beam for reasons not now known and thus can't be seen from far off axis. If not all neutron stars are pulsars, or if old pulsars can not be detected accidentally by radio telescopes that are doing sky surveys, or if the beam has a sharp edge and is not sweeping past earth, then the DV may be a neutron star. The probability that the DV is a neutron star is also set as 0.15, another WAG.

DV as a Black Hole:

A black hole can be as small as the 2.2 solar masses, even if formed during the violent death of a much larger star with an inhomogeneous iron core when it starts to collapse. Conceivable one collapsing core could form two black holes, but they would rotate about their common mass center so quickly that gravitational waves (assuming these are as real as most experts think) would carry away energy and presumably make them spin more slowly and eventually fall together to become one. Perhaps they would radiate away so much mass (Energy is mass, via Einstein's famous equation.) that even two could end up as one with 2.2 solar masses.

If one allows the possibility that the smaller mass dead gray dwarf could become "fat" during the eons it has wandered though space to become the FGD discussed in the first class subsection, then surely the initially more massive black hole should also become still more massive by the same process. Thus if the DV is a black hole derived from a star, it is a young black hole. The supernova of 1006 may have produced a black hole, but it is probably too young. That black hole would need an excessively high velocity to arrive near our sun only a 1000 years later than the light. But obviously there were many older, but still young, "black-hole-forming" supernova in our neighborhood of the galaxy that were either not noticed by man or his state of evolutions was not advanced enough for him to leave any surviving record of the event. Thus if the DV is a stellar black hole, it is a relatively young one, formed by a "neighboring" star of an earlier generation, probably the second of a dead star pair to come our way.—See discussion in prior chapter.

Not all black holes are of stellar origin. Stephen Hawking and others think that in the very early history of the universe, when the energy/mass was concentrated in a small volume that primordial black holes formed spontaneously out of density fluctuations, much like stars would do much later. Black holes do not need to be massive, as the public seems to think. If only one pound of mass were concentrated at a point, it would be a black hole according to classical physics. That is some very small sphere surrounding this point would have gravity so strong that light could not escape for within this sphere, but it is quantum reality, not classical physic theory, that is true, so one-pound black holes may be impossible. (I don't know—this is too complex a question for me.) However, the experts that do know, think some of these "big bang" or primordial black holes even had masses much less than the earth and, according to their theory, are not entirely "black." They also think these small initial black holes can lose their mass, by causing energy to be radiated away, but it also seems possible to me that some could swallow others or near by mass and grow. (I, not Hawking et. al., am suggesting this alternative to their disappearance.) If this is so, then it seems possible that the DV could be a primordial black hole. Perhaps the very small ones do, like old soldiers, just fade away but some may be big enough to resist this fate and compensate for their tendency to lose mass by accreting mass from their environment. It is possible that the DV is one of these, that has managed to slowly grow to 2.2 solar masses as it wanders through space. Thus if the DV is a black hole it is either older than the first stars or a relative young one, born in the death spasm of an inhomogeneous star or star pair within our galaxy (the first member of the pair passing the solar system during the 1920s).

Black Hole Summary:

The main argument supporting the idea that the DV is a black hole is that it is absolutely black and thus undetectable except by its gravitational effect upon other bodies (such as upon PC). There is no strong argument against black holes, only the fact that they are less common than either of the two other candidate classes discussed above, but we are evaluating a single event, not a set of similar events, so I think it is the strongest candidate. The probability that the DV is a black hole is approximately 0.35.

Known Unknowns:

There are many instances in the history of science where firm theory has advance beyond observation to confidently predict unknown objects that were subsequently discovered. Many holes in the periodic table of the elements were not only predicted before they were discovered, but even many of their physical characteristic were predicted as well. (chemical properties, melting points, density etc.) The same is true for elementary particles. Almost all of the fundamental types of matter were predicted by theory and often the next generation of accelerators were designed to have just enough energy to permit the confirmation of these predictions. Such things that are known to be possible in nature usually do exist but merely are currently undiscovered. I. e. they are "Known Unknowns." Two out standing examples exist in this class that might be the DV.

The first is "dark matter." Just as each of the planets orbiting about our sun must move with a definite speed if they are to say in their orbits, so should the stars orbiting about the galaxy center. One can measure this speed and invert the calculation to estimate the mass that is inside their orbit radius causing them to turn. It is found that stars relative close to the center are going much too fast, as if there is some additional mass concentrated near the center. (This is the evidence for a massive black hole at the center of our galaxy). But the stars half way out are going about the same speed and this requires additional mass more uniformly distributed. On this large scale, other stars closer to the center provide part of the additional mass but not nearly enough. Other measurements indicate that all of space has some unseen matter filling it. It is now widely accepted that the stars and galaxies we can see represent only about 10% of the mass of the universe. Ninety percent is some unknown dark matter. Perhaps the DV is a small clump of this unknown matter.

This is not very consistent with other properties dark matter is thought to have. On the scale of the universe, dark matter seems to have an anti-gravity property and be responsible for a major part of the expansion. If this is true, how could it "clump" together to form the DV? The answer to this question is simple, but pure speculation (as far as I know). The force between dark matter could be the inverse of the nuclear force but on a much larger scale. That is, the short-range region of the force could be attractive and the long ranges repulsive. Nuclear forces are repulsive at their smallest scale (This keeps all matter for disappearing into points.) and attractive at longer ranges (This keeps helium and all the heavier element nuclei together despite the electrostatic repulsive force between two or more positive protons that is trying to blow the nucleus apart.) If this is the case and no third region of the force exists, then once the dark matter condenses enough for the attractive forces to dominate, then it disappears into a point. I.e. in addition to primordial black holes, and stellar black holes, there may be small dark-matter black holes. This argument is getting a little too speculative to be firmly in the known unknown class. Perhaps dark-matter black holes should in the Unknown Unknown class discussed at the end of this chapter.

The other alternative "known unknown" is a crystal of alternate polarity magnetic monopoles, probably arranged in a cubic lattice, like common table salt. Maxwell's equations describe all of observed properties of electricity and magnetism but they are not symmetric as they are usually written. That is, they contain a term that explicitly recognizes the existence of electric charge and this term can be either positive or negative. They can be rewritten in a perfectly symmetric form where there is also a term that explicitly recognizes the existence of point magnetic charges, also of either sign. Thus in principle, a north magnetic pole could exist without any associated south magnetic pole, just as a positive charge can exist without any associated negative one. If these equations are made symmetric, an equally valid way to describe the real world mathematically, the speed of light and the minimum electric charge unit (which comes from quantization in nature, not Maxwell's equations) leads to a prediction about the mass of these isolated units of magnetic charge. They are very heavy compared to the electron, but not as heavy as a flea, so the DV is not a single isolated magnetic charge. An aggregate of these monopoles is required if the DV is to be made of magnetic monopoles. They can only form an aggregate if both types are intermixed.

Representing the north magnetic monopole by N and the south by S the following cubic crystal (illustrated in only two dimensions here but extending in three dimensions) should be stable just as a similar array of positive and negative ions in common table salt is stable (although the repulsed force which prevents collapse to a point would be of a different nature.) It would be nearly round, without the corners illustrated here ("cubic" refers to the crystal type group, not the shape of the crystal.)

```
NSNSNSNS
SNSNSNSN
NSNSNSNS
SNSNSNSN
```

There is no known repulsive force between monopoles of opposite magnetic charge to prevent collapse, but it may exist. If it does not, and there is no reason to postulate it does, then this crystal would tend to keep the poles in their same relative positions, but shrink to zero size. That is, there may be a fourth source of black holes, where magnetism rather than gravity is the dominate force causing the collapse to a single point, but away from this point only its gravity would be felt.

There could also be critical size. I.e. for magnetic monopole crystal less than this critical size a modest repulsive force might resist the magnetic compression but then yield to the accumulating magnetic attraction between neighbors (and their gravity also) in the crystal to collapse to a single point. The details need not concern us. The point is that the known laws of nature permit such a thing to exist, and when this is the case, it usually does.

There have been several systematic searches for the magnetic monopole. It is too heavy to be created in current generation particle accelerators that have created many much stranger things. Perhaps all that were created in the big bang condensation of energy into matter have aggregated into dense small crystals or black-hole points at this stage of the universe's development and this is the reason why none have yet been found. The DV may be one such aggregate or point.

Most of the experimental searches for magnetic monopoles have used very intense magnetic fields and tried to pull the postulated magnetic monopole out of some material, typically iron, especially the iron of an iron meteorite that has wandered through space. One Chicago based attempt tried to pull a magnetic monopole out of oysters on the theory that these filter feeders had already processed large volumes of sea water and might have collected one, but personally I think the fact they taste good and are expensive in Chicago had something to do with their selection as one of the target materials tested.

Known Unknowns Summary:

An aggregate of either dark matter or magnetic monopoles has the advantage, especially if they have collapsed into a black-hole point, that they would not be detectable except through their gravitational effects. The argument against both is also the same. Neither has been observed to exist, only predicted by theory. The probability that the DV is a "Known Unknown" is set to 0.15, definitely a WAG. Although either suggested here could exist without collapsing into a single point, it seems more likely for both that they would. Thus taken together all four of the different types of black holes (all five if you want to distinguish stellar cores that collect interstellar matter and "calmly squash" their neutron cores down after a nova from the more violent collapse of an iron-core star in a supernova that leaves a black-hole residue) have a cumulative probability equal to 0.5 of being the DV.

Unknown Unknowns:

As the title of this subsection suggests, not much can be said either for or against the DV being something even creative theoreticians have yet to dream up. Thus I move directly to the summary.

Unknown Unknowns Summary:

Since this is the "none of the above" category and the probabilities must add to unity, the probability that the DV is an "Unknown Unknown" is 0.2, a value higher than anything except a black hole. The strongest argument against this is the fact that no such thing is known to exist, even in theory. Shakespeare put the strongest argument for this class into the mouth of Hamlet. As the history of astronomy has repeatedly shown, there are indeed "more things in the heavens than are dreamt of in our philosophy," but each age is egotistical and thinks it knows just about all there is. Consequently, as a product of my age, I reserved only 20 percent probability for this potentially enormous class.

Chapter 9

Computational Model

Methodology:

The three-body model uses finite time steps and the equations of motion to calculate where each of the three bodies is at the end of the time step given their locations and velocity components at the beginning of the time step. It does this twice for each time step. The first time produces an approximation of the location at the end of the time step. This approximation is based upon the forces acting upon the bodies in their positions at the start of the time step. Using these nearly correct locations for the three bodies at the end of the time step, the forces acting on the three bodies are again evaluated (at this new location). These "end position forces" are averaged with the "start position forces" used in the determination of the first approximation of the end location. This average, composed of the force acting at the start and "end" positions, produces a good estimate of the forces acting during the time step. With this more representative evaluation of the forces acting <u>during</u> the time step, a more accurate determination of the end position and velocities is possible.

The three bodies are then returned to their initial locations and the time-step calculations to a new end location is repeated for the second time, but using now these force that act during the time step, instead of only the start-position forces. The final positions so computed and the new velocity components acquired in this second version of the time step become the new start conditions for the next time step. Prior to beginning the calculation for the next time step, however, the Cartesian coordinates are translated to return the sun to the origin again. This is done so that when the results of all the time steps are completed the sun is still at the origin and the data can be plotted relative to the sun easily. (Sun is always at

$X_s = Y_s = 0$ when the time step starts because I change my coordinate system after each time step to make this true.)

Obviously what is called the final position of the time step could be taken as a second approximation for the end position. The forces could be reevaluated again and a slightly better value for the end position would be achieved but this would represent a 50% increase in computer time and code for very little change in results. This was not done for two principal reasons. (1) With the same computational time as this 50% increase, one can take instead take more smaller times steps and achieve better results. That is, once a good approximation for the forces acting <u>during</u> the time step is available it is better to use time steps that are only 2/3 as large and take 50% more of them to arrive at the same final time conditions more accurately. With approximately circular orbits the error with time-step size grows much faster than linearly. Graphs in the next subsection illustrate this fact. (2) To save computer-coding time and to avoid logical control errors while doing it, a spreadsheet was employed. Even with only two sets of calculations for each time step, more than 50 columns were required and very many pages just to follow one earth year. It becomes too confusing to check for accidental errors if one must look at many different columns to check the entry in any particular box of the spreadsheet. Thus more repetitions of the same two calculation of each time step is not only more accurate, but also presents less opportunity for mistakes than 50% more calculations in each row of the spreadsheet for each time step.

The automatic graphing capabilities associated with the spreadsheet data are extremely helpful in finding the presence of an error. For example, one can set the mass of the DV body to zero and display the <u>closed</u> elliptical orbit of the planet graphically to check that that part of the calculation is working correctly. Finding coding errors in conventionally written programs is much harder and usually some small errors go undetected in most programs, but no small errors could be tolerated in this project as we are trying to fit the DV's parameters to reproduce small perturbations.

Accuracy:

There are three types of potential errors. Modeling errors, Programming errors, and Computational errors. The model is based on the equation $F = GMm/d^2 = ma$. That is, the acceleration "a" of the object with mass "m" by object of mass "M" is GM/d^2 where "G" is the universal gravitational constant and "d" is the distance between objects M and m. This equation does not include the relativistic increase of mass with speed or any other relativistic effects such as the fact that the duration of the time step is not the same for objects moving at different speeds. This is a modeling error, but very small as numerically demonstrated in the next two paragraphs, which the average reader may want to skip.

The fastest object in the center-of-mass reference frame is the planet when it has maximum velocity component towards the DV because the DV is the most massive of the three objects. That is, the sun is "falling" towards the DV faster than the DV is "falling" towards the sun. The planet, annually orbiting around the sun, is moving even faster towards the DV for half of its year, except in the unusual case with the orientation of its orbit perpendicular to the direction towards the DV. In that unusual case the planet and the sun have the same velocity of fall towards the DV. Thus the highest speed object in the simulation is the planet.

Because it is closest to the sun, earth is the planet moving fastest with respect to the sun of all the planets separately considered. (The DV's effect upon Mercury and Venus was not investigated. They are tightly bound to the sun and it is mainly earth I was concerned about. If Earth's orbit is only slightly changed, as turned out to be the case, I knew neither Mercury nor Venus would be disturbed significantly enough to be important to us. The outer planets, however, might be thrown into elliptical orbits and hit the earth so each was investigated.) Earth's initial orbital speed is 0.0007168 AU/ hour. The best-fit mass of the DV is 2.2 times the mass of the sun. Thus the sun is falling towards the DV 2.2 times faster than the DV is moving to the sun, as the center of mass is stationary. The best fit for the DV's initial speed is $V_x = 0.0016$ AU/ hour and $V_y = 0.0039$ AU/ hour. (By definition of the coordinate system $V_z = 0$.) Consequently the DV's speed, relative to a stationary sun, is 0.004215 AU/ hour (approximately six time earth's orbital speed). Thus the sun's speed is 0.009274 AU/hour and the earth's speed could be 0.01 AU/hour (0.001768 + 0.009274 = 0.01). Actually it could be higher as DV and sun are accelerating as they approach each other but the initial speeds will serve to evaluate the significance of relativistic effects. The speed of light is 7.2 AU/ hour so the "v/c" ratio is only 0.01 / 7.2 and relativistic effects go as the square of this ratio. Consequently the neglect of relativity is an error on the order of two parts in a million and thus fully justified. (Both sun and the DV also appear to have additional speed relative to other stars as does our local group of stars, but these speeds are not important for this question as our Cartesian reference frame also has these speeds.)

In the case with the planet being earth, a circular orbit approximation was used. This is justified to avoid the need to compute and enter six values (exact x, y and z speeds and x, y, z location) for each starting point considered as doing so presents the opportunity for manual entry errors. Instead only three angle coordinates were given to specify earth's initial location. (With this approximation and astronomical units, the initial orbit radius is unity.) This is a modeling error, much larger than the neglect of relativity effects, but still not considered significant because the main effort is directed towards determining the DV parameters as accurately as possible and earth has little effect upon DV compared to the sun's forces. This error is more significant when the effect of the DV upon earth is considered, but not as large as that associated with the uncertainty in the best-fit parameters of the DV. Also there is considerable uncertainty in prediction the climate given any set of values for the new eccentricity and orientation of the major axis of the ellipse with relation to spin vector of the earth. (This relationship determines where the earth is in its orbit during each of the seasons.) Thus to predict the climatic effects, highly precise values for these new orbit parameters are pointless.

In the case of Pluto the use of six initial staring condition parameters was unavoidable because of its orbit eccentricity is much larger. When Pluto was the planet, the angular coordinates and radius values normally entered were still supplied and the automatic calculation of the corresponding position and three speed components (in the circular model of the program) served as a rough check upon the manual entries that "over rode" these automatically computed values. That is, in the spreadsheet the formulae that convert these angular inputs into xyz location and speed components reside in one row of the spreadsheet and the cells immediately below these values are simple copies whose values can be replaced without destroying the formulae. The time-step model's starting values for the initial calculation are the values found in these second or lower cells, which in Pluto's case were manually entered.

The final modeling error (known) is the assumption that all of the motion of the sun and the DV is confined to a plane, the X, Y plane of the coordinate system. (Next chapter discusses the coordinate system in detail, but most readers will want to skip that entire chapter.) That is, any effect of Jupiter and the other planets to make the DV and sun "scattering" interaction into a multibody three dimensional problem was ignored, including the "planet's" forces in the Z direction upon both the sun and the DV. Obviously their Z forces upon the planet were included but the orbit plane of the planet remained surprising little changed. This is because the small tilt produced while the DV was approaching from one side of the orbit plane was largely cancelled out during its retreat path on the other side. Thus the earth's rotation axis, whose orientation in space is strictly unchanged, remained with essentially the current 23.5-degree inclination to the ecliptic.

Great care was exercised to avoid programming errors. As already mentioned, the logical control of computation flow was left to the spreadsheet, rather than written uniquely for this problem. Also as already mentioned, graphical testing of intermediate results was extensive used and extreme values of input parameters were used to make known results. The example, already given of tests with DV mass = 0, illustrated this concept. As a second example, the planet can be artificially given a large mass and only +X velocity initially. Then the model must predict that the planet's trajectory graph is nearly a straight line, parallel to the X axis, as the planet leaves the sun in an XY plot of its position. Etc.

Multiple methods of computing were also used whenever possible and their results compared to be sure they were the same. For example in Pluto's case, it orbital speed varies significantly around the orbit. One approach to computing it is to compute the area swept in each time step (a constant when the DV mass is set to zero) and confirm that the current distance from the sun times the speed component perpendicular to it is a constant. I.e., approximately the shape of the area swept in a small time step as a triangle with the sun at the acute apex and base proportional to that component of the speed. Alternatively one can use the fact that the speed ratio of a planet in any two different locations of the orbit about the sun, even widely separated ones, is the same as the inverse ratio of the sun's distances from their respective tangent lines—See Newton's *Principles of Mathematics*, Book I, Proposition 16, Theorem 8. These are fundamentally the same calculation but the formulation is very different so a mistake in the formulation of either calculation will cause the results to disagree.

Some computational errors are unavoidable as only a finite numerical accuracy is carried in each time step. Thus extremely small time steps can essentially eliminate the modeling errors due to finite time steps but as each new position is based on the terminal position and velocity of the prior time step too many small time steps can compound the rounding errors in the numerical representations of these conditions. Fortunately these types of errors tend to average out. At times it is possible to compute values needed in each new time step directly from the original starting conditions. For example, with the DV mass equal to zero for testing, it is really a two-body problem with an analytic solution. Thus Newton's Proposition 16 Theorem 8 can then be used to calculate the starting condition speeds of each time step directly from the original initial speed and position (one numerical truncation error) rather than a series of small but compounding errors when the starting speed of each new time step is computed from the accelerations and the prior time-step's speed. Unfortunately, in the situations of most interest, no analytic solution is possible and one must base the next step's computation on the results of the prior step, but one can confirm that numerical truncation errors are not significant by comparing these "step-by-step" results with the "direct-from-starting-condition" results in these special cases and be confident that the numerical rounding errors are not significant in the more general case.

The change in the results with different time-step sizes is also a useful tool for evaluating the over all accuracy and this can be use when the DV has realistic mass (not set to zero for testing). For example, after the DV has gone far past the point of closest approach, the perturbation is over and the new orbit should continue to repeat without change. If it does not, some error is present; usually the time step is too large. Three graphs for the earth's result with different time steps are now presented on the next three pages to show this test. In the first graph the time step is set equal to 6 days, which is 50% larger than the similar graphs presented later, (*in Chapter 12*), yet the time-step error is acceptable small, only about 1% total during the last five years shown on the graph. (0.2% per year) This is in large measure due to the use of two calculations for each time step. (Second calculation based on forces acting during the time step as discussed earlier.)

The second graph has time step of 9 days, and even though the DV is very far away when it ends, the earth's computed distance from the sun is still decreasing slightly each year because the time step is too large. In the final twelve years of this graph, the maximum distance from the sun has decreased from 1.06 to 0.92 AU. (Slightly more than 1% per year.) Thus a 50% increase in the time step has produced more than a 500% increase in the error rate. These two results illustrate the non-linearity I mentioned earlier, and the next graph shows it fully.

The final graph of this series has a time step of 12 days. Although only one third larger time step than the 9 day graph, that had a very modest error each year, this 12 day graph "explodes" with the earth being hurled into space in the last few time steps after day 4000. Note that the calculations with a 9 day time step are still successfully reflecting small oscillations in the earth's separation from the sun due its eccentric orbit at 6800 days, but the 12-day time-step graph is totally erroneous by day 4000. The point is that the error with time step is a strongly non-linear function of the time-step size. The fact that there is insignificant error with the six-day time step is thus proof that the four-day time step routinely used is essentially free of this modeling error. See similar four-day graphs at the end of my report. (*Now in Chapter 12 as I have added an appendix to Jack's report that gives some of his computer code and early comments. Appendix A1 is a historical document of considerable significance. It marks the beginning of the ADV era.*)

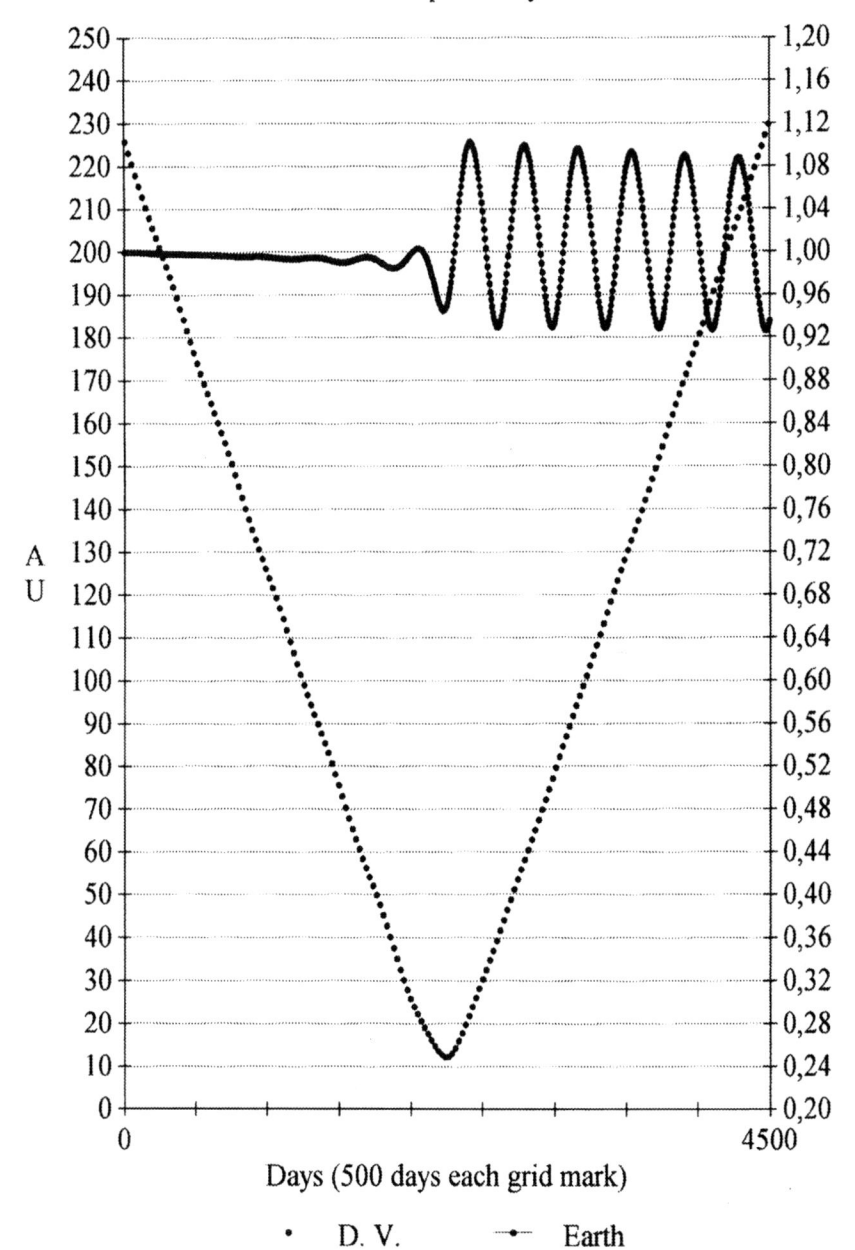

Distance to Sun
Time Step = 6 Days

A U

Days (500 days each grid mark)

• D. V.   •— Earth

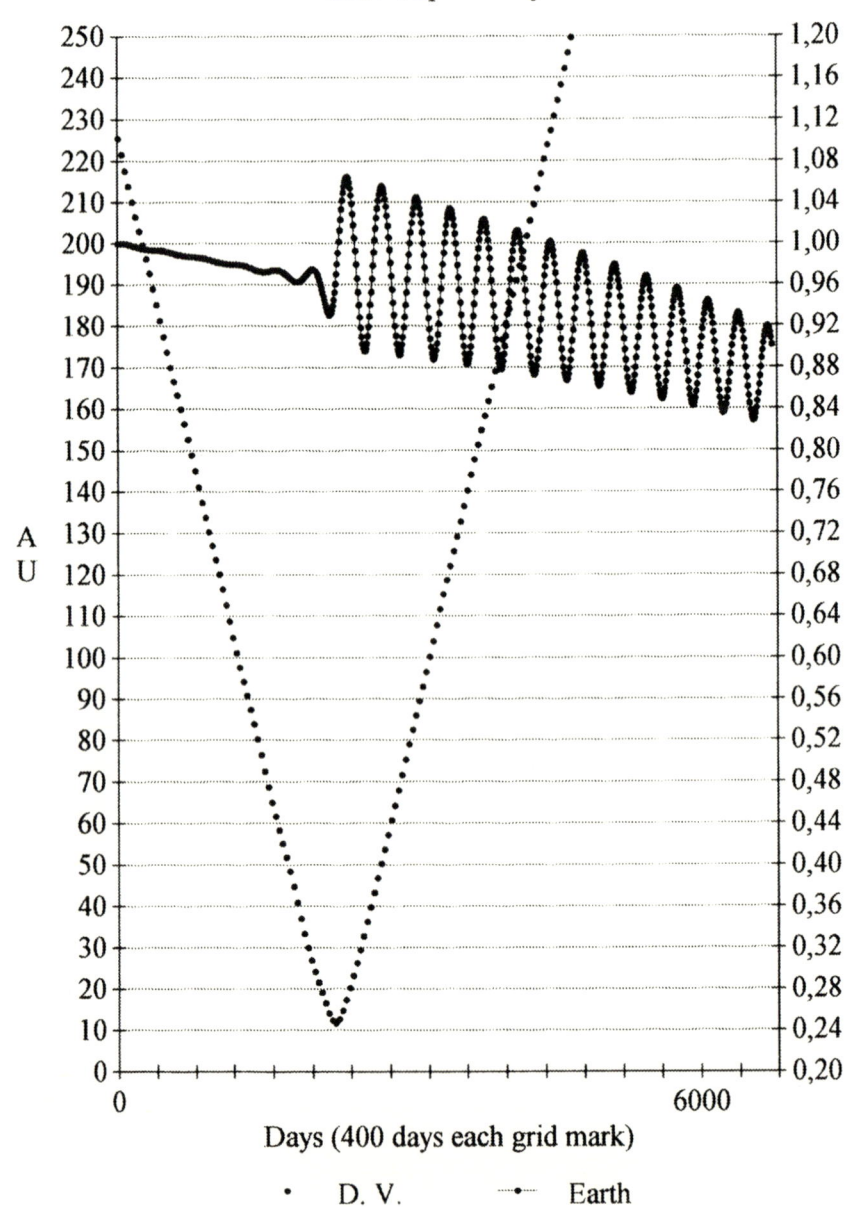

Distance to Sun
Time Step = 9 Days

Days (400 days each grid mark)

• D. V.   —•— Earth

Distance to Sun
Time Step = 12 Days

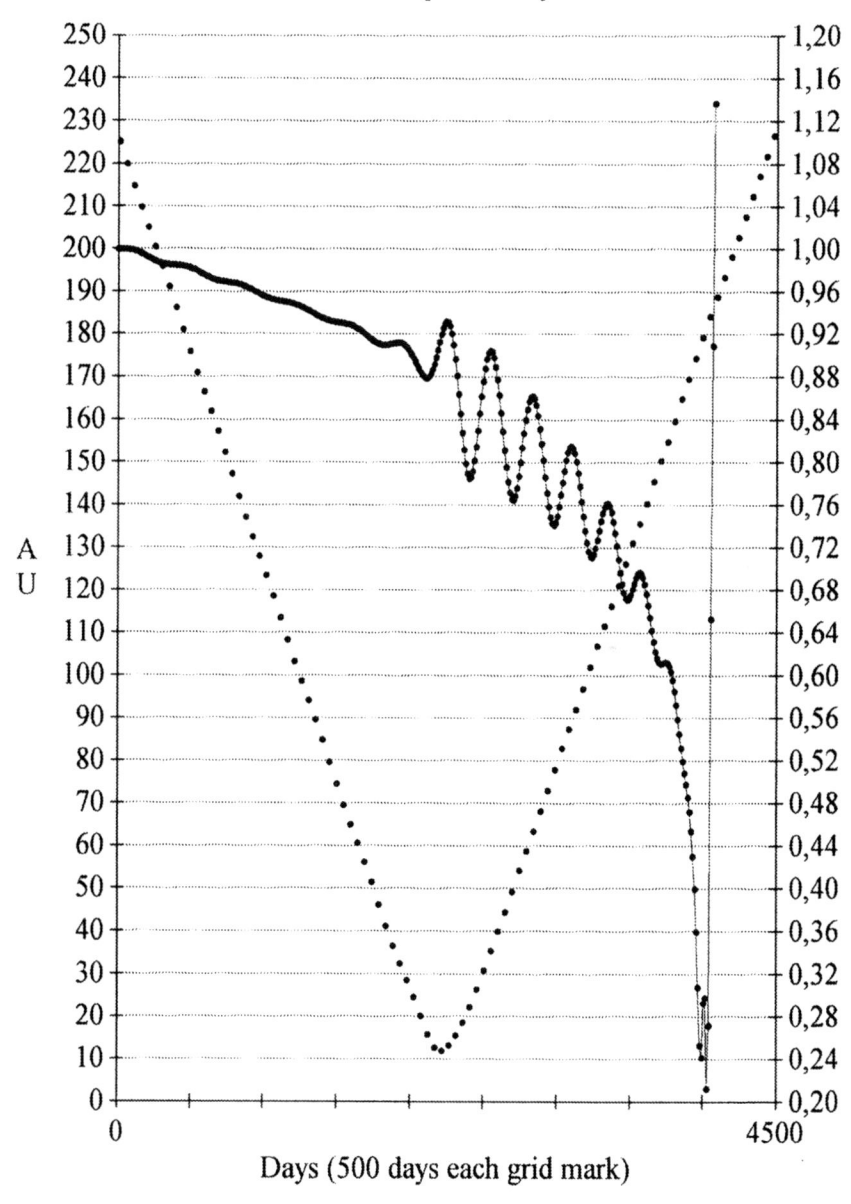

Obsolete Versions:

Initially, because the mutual scattering of the sun and DV is essentially confined to a single plane and already known to result in open parabolic trajectories for both, these facts were assumed. Any two-body central force problem can be reduced mathematically to that of a single body scattering off a fixed point but the mass of the scattered body is reduced to $Mm / (M + m)$ where M and m are the masses of the original two bodies. Thus the DV was assigned a parabolic trajectory characterized by two parameters and moved along it in steps without the need to calculate the forces that were acting on it. Later this approach was abandoned, mainly because of doubts relating to whether the mass that should be acting upon the planet was the true mass of the DV (M) or the "reduced mass." Also some of the other assumption built into model about how the sun should be treated became questionable during its use.

Postscript: The first pages in the spreadsheet implementation of that model are included in the appendix, at the instance of my historian friend assembling this book, because unlike the final three-body model used, it has many notes and comments written during the initial period when the existence of the DV was first postulated. He seems to think someone might be interested in my state of mind and ideas during this early stage of the effort. I doubt this, but agreed to include it, even though I am a little embarrassed by it now. I want the reader to know it was put together quickly in a few long days, when I was anxious to have some idea of the characteristics that a dark visitor would need to have to make the observed perturbations. What is important is that the DV exists and is soon going to be changing life on earth, not how or who discovered this first, but I would like the credit I am due for suggesting it as the cause of the previously unexplained perturbations. I will drop my pseudonym (Jack) in a couple of years when the initial shock has dissipated. Only I have the dozens of glass-plate photographic observations recording Pluto's precise position, on which the calculation of the DV's trajectory is based, so it will be easy to prove my identity when I "come forward" if it is ever questioned.

Billy T, also a pseudonym, is the most idealistic man I know. I don't think even the CIA would be able to make him identify me. I hope the reader will not think me too paranoid when I mention that the glass plates, with the precise positional data, have been stored in such a way that if anyone but me tried to open the safe, they will be shattered into thousands of pieces and dust by a small explosive. I mention this fact for my own protection, just incase Billy T can be found and made to talk by some chemical injection. "George, Mr. Goldwater" etc. are not their real names either. I was not married in Boston, etc. but I did go to Harvard, as did many others. "Karen" went to a different school, other than Radcliffe, in another city, where Billy T was an undergraduate student, and we did not meet in the "Green Mountain Estate." Billy T's history is essentially true, but almost all the names and places given are not the actual ones. Even his detailed description of my Ph.D. project is only one thing a physic student I knew was considered doing, without any relationship to what either he or I actually did. My father did not own a cattle ranch, but was wealthy for other reasons, etc. The whole story presented is one of "factual substitution," with invented facts, that closely resemble the true history, replacing the actual ones. All that is really true exactly as stated is that I went to Harvard, own a small private observatory in South America, and that the DV is coming. Only Billy T, who has known me for years, could do the work he has done so well without constantly bothering me. I thank him from the bottom of my heart and will see that he is properly rewarded. (Billy, you are not to change a word of this paragraph. I want the full truth told, as much as we can, now.)

Chapter 10

Coordinates and Initial Values

Note: Most readers can skip this chapter. I provide it mainly for other astronomers and readers who may want to construct their own three-body model of the dynamic interactions; however, the footnote at the end may interest all. (*See also Appendix A2.*)

Coordinate System:

The origin of the Cartesian coordinate system used in the computational model is fixed in the center of the sun. The X and Y-axis are in the DV / sun scattering plane. That is, in the plane in which all the DV and sun's motion occurs. By this definition, neither has any Z-axis motion. This is possible, as discussed in the prior chapter on model assumption and accuracy, because the tendency of the planets, especially Jupiter, to curve or warp the scattering "plane" is neglected. The "Jupiter effect" will be evaluated when my four-body model is completed and "de-bugged." It should be very small while the DV is far from Jupiter and thus not significantly change the "best-fit" results already achieved with the three-body analysis model.

In the computational model only the planet is subjected to forces in the Z-axis direction. Recall also from the prior chapter's discussion of the computational model that after the end of each time step, the coordinate system is changed (pure translation, no rotation) to move the sun back to the origin of a new coordinate system despite its gravitational acceleration towards the DV during the prior time step. Thus the sequential set of coordinate systems used during the analysis all have their origin fixed in the sun.

By a convenient choice, the X-axis is the line of intersection between the scattering plane and the earth's orbital plane. That is the X-axis is in both the ecliptic and in the scattering plane. The +X axis passes from the center of the sun though a celestial point which has a Right Ascension of approximately 150 degrees, RA = 10 ± 1 hr and has a declination of approximately 12 ± 4 degrees North. (I am not now giving precise values—mainly for reasons discussed in the next paragraph.) The +Y axis of the DV / sun scattering plane is in the Northern Hemisphere in the polar region, approximately 10 ± 5 degrees from the celestial pole.

The location of this plane is known much more accurately than given in the prior paragraph, but the "best fit" celestial description of its location is not currently being divulged as I hope to be the first to observe the DV, if optically observation is possible. Even if it is not possible to observe the DV optically, as it passes essentially in front of background stars, its gravitational distortion of space will "wiggle" these stars. These observations will let me refine my current knowledge of its trajectory, but I don't want others to do this and publish first. Thus I am keeping the best-fit data secret still. I do, however, feel the public has a right to know the DV is coming. Hence I asked my trusted friend, "Billy T," to construct this book. I am too busy with my observations and searching for "wiggling stars" with my flicker table, to do this now.

The +Z axis, which is normal to this XY plane, is also in the Northern Hemisphere with RA = 4 ± 1 hr where the northern declination of the ecliptic is relative large (approximately 21 degrees North). Thus, even thought the +Z axis normal of the XY plane is tilted above the equator 10 ± 5 degrees into the Northern Hemisphere in celestial coordinates, near RA= 4hr, the northern declination of the ecliptic is greater. That is, the Northern Hemisphere normal to the XY plane is on the opposite side of the earth's orbit plane from the Northern Hemisphere normal of the ecliptic. The sign convention adopted for angle between the XY plane and the planet's plane is that if both Northern Hemisphere normals (of the XY plane and the planet's plane) are on the same side of the planet's plane, then the angle between these planes is considered to be positive. With this sign convention, only Pluto has a positive angle between its orbital plane and the XY plane. In descriptions of the computation model this angle between planes is designated "OPx" where x designates the planet. (In the actual model, it is just OP as there is only one planet in the current three-body model.) It is referred to as the Orbit Plane angle. The earth's angle OPe is negative 79 ± 5 degrees. That is, the ecliptic and the scattering plane are almost perpendicular to each other.

Despite the use of orthogonal Cartesian coordinates in the analytical model the initial location of the planet is input to the model in terms of three angles and its distance from the sun. The first of these three angles is the angle Orbit Plane angle, OP, just discussed. Note -90 ≤ OP ≤ 90 degrees. A sign convention is required because two entirely different planes, both with the same intersection line, can differ from being perpendicular to the XY plane by the same amount. For example, there are two different planes that are 11 degrees off the perpendicular to the XY plane and have the same intersection line, e.g. RA = 10 hours.

The second angle is the angle between the +X axis and the Intersection Line, the line in both the XY plane and the planet's orbit plane. This angle is designated ILx in the analytical model descriptions. By our definition of the +X axis, for earth, ILe = 0 but for Pluto, ILp = 9 degrees approximately. IL is a measure of the angular distance between the intersection line and the +X axis. The sign convention adopted for angle IL is that IL is positive if the intersection line is in the first quadrant of the XY plane, That is, if the intersection line passes between the +X and +Y axes. IL is negative if the intersection line passes through the fourth quadrant of the XY plane. Thus IL is bounded by -90 ≤ IL ≤ 90.

The third angle specifies where the planet is initially (at the start of the time-step model) in its orbit about the sun. The Angle of the Planet, AP, specifies how far the rotation about the sun has carried the planet from the +X axis. When the planet passes through the intersection line, AP is either 0 or 180 degrees. The sign convention adopted is that zero corresponds to a place on the intersection line with positive value for its X-axis coordinate. AP is bounded by 0 ≤ AP ≤ 360.

Initial values:

I can not give both the starting date for the earth's graphical results that I present later and accurate values of AP as this would accurately specify the declination of the +X-axis as this is in the earth's ecliptic. If the +X axis declination were known accurately then the RA location of the XY plane would be accurately reveled. (Each declination of the ecliptic corresponds to two RAs, 12 hours apart, but I have already specified approximately which.) The reader is surely more interested in when things are going to happen to the earth rather than what angular values were input to the computational model. Thus I chose to tell that all graphs begin with the earth near its autumnal equinox position of the year 2003. Most display the next 3000 days and thus end a few days after the start of year 2012, but by then a new calendar may be in effect as the new orbit of the earth is 378 days for each year.

I can, however, give the Dark Visitor's "best fit" speed and location in this somewhat vaguely specified XY plane as projected for 23 September 2003. Then the DV will be at X = 97 AU and Y = 204 AU. This is essentially 226AU from both the sun and from the earth because the DV is coming from the polar region (nearly perpendicular to the plane of the elliptic). It is traveling fast, Vx = -0.0016

AU/hour and Vy = -0.0039 AU/ hour projected for 23 September 2003, and speeding up as it "falls" towards the sun. Thus its speed in the XY coordinate system on that date is projected to be a little more than 0.0042 AU/ hour. It also appears to have essentially the speed of the sun in the local group. The "best fit" for the DV mass is 2.2 solar masses; thus it is more accurate to say the sun and earth are "falling" towards it at almost 0.0093 AU/ hour or that the DV is closing the earth at approximately 0.0139 AU/hour. The DV will pass the earth (at a distance of 12 AU) in November of 2009, if the current calendar is still in use.[15]

---

[15] My friend, the historian integrating this report with his historical account (and George's projections of the climatic impact of the DV), and I have a bet about this. I have great faith in human inertia (and the inability of nations to agree) so I claim that we will still be using the old calendar on 1 January 2012, but he thinks that will be approximately the start of year 3 ADV. He always seems ready to throw over the old ways, but I doubt he will leave this comment in the book.—*Your wrong Jack—I am proud to be willing to change when it is necessary and desirable!*

CHAPTER 11

COMPUTATIONAL RESULTS

General comments:

The results of the three body computational model are presented graphically in the next chapter. The first graph for the earth plots the z and y-coordinate variations against the x-coordinate variation. The fact that the y-variation does not deviate much from a line through the center of the more circular curve does not imply any highly elliptical orbit, only that the plane of the orbit is nearly the same as the XZ plane of the coordinate system. Each dot is four days later than the preceding dot (or earlier if you go in reverse). You can easily count the number of intervals between dots on the nearly circular curve. (You should find 94 intervals with a small gap extra at the top because not quite a complete orbit is plotted. The gap is nearly two days.) Thus the new year is almost $4 \times 94 + 2 = 378$ days long. Notice also that the lower half of the z curve nearly dips beyond negative 1.1 but the top half, where the small gap is, does not reach even to 1.0. Later graphs give more exact values for these extremes in terms of the total separation form the sun, not just one component of the separation. Notice also that the dots are closer together at the bottom of the "circle" in comparison to their spacing at the sides or top. This feature will also be discussed in more detail with later graphs.

All the other graphs are similar to each other in that they use time as the horizontal axis, rather than one of the coordinate variations as the first graph employed. The left-side scale of these other graphs (usually ranging from 0 to 240 AU) gives the separation between the dark visitor and the planet in astronomical units. The plot which uses this left side scale is a roughly "V" shaped series of unconnected dots, each dot being four time steps later than the preceding dot. The time-step size used in the calculation is given as a subtitle at the top of the graph. For example, when the time step is four days the interval between the dots representing the separation between the dark visitor and the planet is 16 days.

I only give graphs for earth and Pluto. Pluto's graph does not give the x, y and z-coordinates separately but it has the same starting date as the earth's graphs. Unfortunately, I can not present graphs for other planets or give the x, y and z-coordinates separately for both earth and Pluto and still preserve some uncertainty about the "best-fit" location of the dark visitor as a function of time.[16] This omission is of little concern to the average reader because for Pluto, these omitted curves change slowly during one earth year.

The vertical grid marks of the earth's three graphs are at intervals of 100 or 50 days for the final "expanded" pair. Pluto's first graph has vertical grid marks at 100 day intervals and extends to day 3000, like the earth's first graph. Pluto's last graph has 200-day intervals and extends to 6000 days. All graphs have a legend at the bottom that helps the reader remember these intervals and identify which curve is which variable. All graphs begin with three bodies of the model in their position as of approximately 23 September (the earth's current autumnal equinox) of year 2003 of the current calendar. It is a trivial exercise to calculate where the planets will be on the starting date, from their orbital elements, not their celestial coordinates, because none have yet been significantly displaced by the dark visitor. I can not, however, give the exact location and date for planets in the coordinate system with the XY plane being the scattering plane of the sun and the dark visitor interaction without disclosing precisely the orientation of this scattering plane.

In the case of earth, the graph presents the x, y and z-coordinates of the first year's orbit (still with 365+ days) to help the reader understand the coordinate system discussed in the prior chapter. The z-axis coordinate is continued for all the years of the graph. On the starting date, the earth has recently passed through the negative portion of the X-axis. That is, angle AP, discussed in the last chapter, is greater than 180 degrees but less than 270 degrees so the initial

---

[16] If the DV's distance from three known points were disclosed, even for only one moment of time, its location is accurately known at that time. This is the way the GPS system is used to locate any point in space near the earth or on its surface from the known location of three satellites. (If you know you are on the surface of the earth, only two are often sufficient.) Since these graphs present two distances (from Pluto and from the earth) for many different times, even this limited information can be used to determine the XY-scattering plane more accurately than I would like it to be known immediately, but it is a lot of work and because that plane is nearly perpendicular to the ecliptic, considerable uncertainly still remains.

value of the x-coordinate is negative. Because the initial x-coordinate value can be read rather accurately from the graph, I must give the starting date only approximately in order to keep the exact orientation of the XY plane secret.

I extended the z-axis oscillations across the entire period of the graph as this permits the reader to easily know how many years have passed since the starting date, without dividing the number of days by some number between 365 and 378, the length of the year for the last two years illustrated. Initially the positive peak of the z-coordinate coincides approximately with the end of November and start of December, but this will change as the year becomes longer. What will not change, and thus serves as a more convenient reference, is that the positive peak of the z oscillation marks the beginning of the coldest part of winter in the Northern Hemisphere (and correspondingly the negative z peak marks the warmest part of summer there). I hope that, by giving a full year of the earth's orbit in xyz-coordinates, I have not already divulged too much to other astronomers who have access to larger telescopes than mine. To help insure this is not the case, the small y-axis oscillation displayed in the first two graphs has been modulated by a near unity, but undisclosed, factor.

The right-side scale of the graphs is used for the earth's three coordinates, x, y and x. Because the XY scattering plane is nearly perpendicular to the ecliptic, and the X-axis is in the ecliptic, the x and z- oscillations are essentially in quadrature and both of amplitude nearly unity. (I am aware of the fact that with this information, other astronomers can relatively easily refine the RA of the scattering plane previously given in the text but because of the undisclosed factor of the y-oscillation graph, the declination of the +Y axis in the polar region should remain uncertain and this is where one must look for the DV now.)

The right-side scale is also used for the most important graph presented. That is, the distance between the planet and the sun. For earth in a circular-orbit approximation this curve automatically has a value of unity before the dark visitor perturbs the orbit. For Pluto the natural unperturbed value at the start of the graph would be approximately 31 AU, but the variation displayed would be very small if a scale including 31AU were used to display the x, y and z-coordinates also because, like the earth's x, y and z, these coordinates can be both positive and negative. (Pluto's x, y and z-coordinates were routinely displayed, but for reasons already stated, are not presented.)

Consequently Pluto's unperturbed distance from the sun was "normalized" to unity but not by the unusual process. Instead of dividing by its unperturbed orbital value, as is usually done in a standard normalization, all but one AU of the unperturbed orbit value was subtracted from the perturbed value computed. Thus the deviation from unity is still absolute, not normalized. The next subsection begins with an example using data from the last graph of this chapter (*now this is at the end of Chapter 12.*) to clarify this.

Specific comments, Pluto:

At the end of the 6000 days displayed in Pluto's last graph, the dark visitor has moved Pluto 1.43 AU closer to the sun. 1.43 = 1 - (-.43). By this "offsetting" instead of standard normalization, the display of the dark visitor's effect is enhanced approximately 31 times in the graph and more easily evaluated. It is not possible to lengthen the time step enough to display a full orbit of Pluto for failing accuracy reasons as already discussed and illustrated graphically (*in Chapter 9*). The reader will no doubt be interested to know how close to the sun (and earth) this "falling" Pluto will come. (I certainly was.) Fortunately, here there is good news. At its closest approach to the sun, Pluto will still be approximately 17 AU away, far beyond earth. Its new apogee will be approximately 49 AU. Thus it poses no direct threat to the earth.

Basically what will happen to Pluto is that the dark visitor will pass behind it and its gravitational pull will slow Pluto down, but will not immediately move it much while the DV is quickly passing through our solar system. Lacking some of the kinetic energy it had, Pluto will not be able to resist the pull of the sun's gravity and begins to fall closer to the sun. It still has considerable tangential speed so this fall is not directly towards the sun. As it moves closer to the sun it gains speed. When it is only 17 AU from the sun it is going so fast that it will be able to climb up the sun's gravitational "hill" all the way to 49 AU but then begin its fall again.

Pluto's new period will be approximately 340 years. The old one was 248 years. It will no longer be in a 2 to 3 resonance with Neptune. It has always crossed the orbit of Neptune but because of the resonance and difference in their orbit inclinations, there was never any danger it would scatter off Neptune and threaten earth. After the dark visitor passes it will also cross the orbit of Uranus, but Pluto will still be inclined the to orbit plane of the other planets sufficiently that it will probably never scatter significantly off either Neptune or Uranus. In any case, with a period of 340 years, the opportunity to do so will not arise often.

The new orbit of Pluto was estimated by fitting an ellipse[17] with the sun as the focus through two points of the orbit despite the inability to extend the time step

---

[17] "Billy T" passionately believes that history is not remembered or useful if it is a memorized set of facts. I feel the same way about math and physics. It is pointless to memorize many formulae. In this long footnote I demonstrate a better approach, illustrated by solving the problem of calculating how close Pluto will come to the sun.

sufficiently to compute more than a small part of the orbit. The last position computed (day 6000 where the still very small numerical error was largest but the remaining dark visitor perturbation was least) was combined with the position at day 3000, where the remaining dark visitor perturbation is greater but the numerical error is less. Because Pluto has a small but significant y component to its motion it was necessary to first rotate the coordinate axes to make these two points and the sun all in one plane before computing the ellipse. Once the ellipse is know, the period is simply the square root of the cube of the semi-major axis, in

---

I am confident that all readers of this footnote already know that the equation of a circle is $X^2 + Y^2 = R^2$. This can be written as $(X/R)^2 + (Y/R)^2 = 1$. An ellipse has the same basic equation, but it does not have a constant radius R. Instead when Y is zero, X must equal the "R" of its denominator which is traditionally called "a" and likewise when X is zero, Y must equal its "R" or "b." That is, the equation for an ellipse is: $(X/a)^2 + (Y/b)^2 = 1$. (The same equation also produces a hyperbola if a - instead of + sign is used.) More general forms of the equation can be constructed from this basic one by translation and / or rotation of the coordinate system axes. I don't want this long footnote to be even longer, so I will just illustrate a simple translation. You could, for example, place the center of the ellipse on the X-axis by at the point X = c instead of at the origin if $((X-c) / a)^2 + (Y/b)^2 = 1$.

For Pluto, we suppose that the ellipse must pass through the two point (X1, Y1) and (X2, Y2) that are discussed in the main text. The way you achieve this is to insert the X & Y values of one of these points into the simple ellipse equation and use algebra to solve the resulting equation for either a or b. If you use the first point and solve for a, the result is:
$a^2 = X1^2 / (1 - (Y1/b)^2)$. (We leave it as "a squared" as that is more useful than a.)

Now we place this value of a^2 in the equation $(X/a)^2 + (Y/b)^2 = 1$ and insert the other of the two points and solve for b. When you do this, after a little algebra, you get:
$b^2 = (Y2^2 - (Y1 \times X2 / X1)^2) / (1 - (X2 / X1)^2)$

Now you can compute the value of b as only the coordinates of your two points appear in this equation. Once "b squared" is known, place it in the first equation for "a squared" and learn the value of a. Then both a and b are known and you have the equation of the ellipse passing though Pluto's two points.

Please do not even consider memorizing any of these derived equations. You will just forget them. I only gave them so you could check your own derivation. Learn a few facts and derive what you need, when you need it.

earth years if the length of the semi-major axis of the ellipse is measured in AU. That is:

Period P = Square root (a^3) where "a" is the semi-major axis.

Specific comments, Earth:

In the second graph presented in this chapter (*Chapter 12 now*), the open circles of the x-axis curve that display the first year of the earth's orbit are successive time

---

In our current application we were not primarily interested in the equation of the ellipse that passes through the two selected points. This was just a means to the end of interest. We wanted to know how close Pluto would come to the sun (and earth). The ellipse we have solved for is centered on the origin but the sun is at a focus and now we need to know how far the focus is from the origin.

To solve this new problem, we draw upon our store of useful (and easily remembered) facts about ellipses and their foci. Namely that the total distant to any point on an ellipse from both foci is a constant. (A convenient way to draw and ellipse is to use a string with ends tied to the two points you want to be the foci.) I'll call the length of this string "S" and the distance each focus is from the origin "f" in the figures below. The dotted line represents a string of length S going to two different convenient points on the ellipse. In the left figure it goes to a point "a" from the origin and in the right to a point "b" distant from the origin.

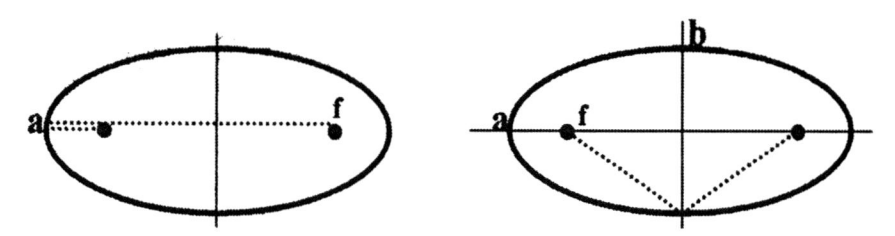

From the left figure we see that the length of that portion of the string going from the right focus to the ellipse is (a + f). From that point on the ellipse back to the other focus is (a-f). Thus one equation for the total length of the string is (a + f) + (a - f) = 2a = S. The right figure gives a different equation for the length of the string, using the Pythagorean theorem, S = 2 x Square root (f ^2 + b^2). Thus (S/2)^2 = a^2 from the first equation and also (S/2)^2 is (f^2 + b^2) from the second. Hence f ^2 = (a^2 - b^2). (I know this is a simple, general equation, but don't memorize it either. Remember how to derive it.)

steps and thus are 4 days apart. The amplitude of the y-axis oscillation is so small that the time-step dots of that curve all run together making that relatively unimportant curve into a fat sine wave. The dots of the most important curve, the distance of the earth from the sun, which appears at the top of the page with initial value of 1 AU have also run together to make a fat complex curve. The area at the right end of this important curve is of the greatest interest as it corresponds to the condition the earth will be in after the dark visitor has left our solar system. Consequently I present two more graphs of this area. In the first one, I expand the right scale to range only between 0.60 to 1.20 AU and present only the last 850 days of the previous graph. Just enough to clearly display the closest approach of the dark visitor. In it you can see that The DV will miss the earth by 12 AU.

The second "expanded graph" displays only the last 500 days with a scale that spans only 0.93 to 1.11. In it, the individual dots, which are four days apart, are everywhere distinct, rather than merged together in the peaks. In it the reader can count the dots to see that the earth will be more than 1 AU away from the sun for

---

Now we can finally answer the question we were interested in, namely: How close will Pluto come to the sun? The answer is (a - f) which we have just shown is: a - square root ($a^2 - b^2$). Thus we know how close Pluto will come to the sun in terms of a and b, but we know a and b in terms of the coordinates of the two chosen points (X1, Y1) and (X2, Y2). So the problem is solved. I took these coordinate values from the numerical results and with the method just outlined found that Pluto will not come closer to the sun than 17 AU or surely miss by at least 15AU as my two points were not entirely accurate.

The real point of this lengthy footnote is to show that a few useful facts and understanding of them is much more useful than hundreds of memorized formulae. What did I really know to solve this problem? Answer: (1) Pluto's two points were on an ellipse; (2) A simple equation for some conic sections; (3) A few simple rules of algebra; (4) The Pythagorean theorem and, (5) All points on an ellipse have the same total distance to the two foci (or that an ellipse can be drawn by a string with ends tied to the foci).

My friend, the historian assembling this book, feels strongly about how history should be taught and applied in daily life. I feel equally strongly that math and physic can be simple and applied in daily life if one tries to understand a few basic ideas and refuses to memorized many formulae. Unfortunately it is not always taught this way. If this footnote was not worth your time, I apologize, but I think if you absorbed the underlying philosophy, it is the most important one in the report. (*End of Long footnote*)

more than "57 dot gaps." The actual time is a little less than 230 = 4x57 + 2 days when one consults the numerical output of the model instead of the graph. The "dot counting" for the full year is easy if one uses the fact that the DV's dot gaps represent four time steps. (23 dot gaps x16 + 2 dot gaps x4 + 2 days = 378 days for the year.) The maximum and minimum distance from the sun can be read accurately from this graph as 1.108 AU and 0.937 AU. This implies that the eccentricity, e, of the new earth orbit will be 0.0836 instead of the current 0.0171.

e = (D - d) / (D + d) where D is the maximum and d is the minimum distances from the sun. For a circular orbit, D = d so the eccentricity of a circular orbit is zero.

The semi-major axis is thus (1.108 + 0.937) / 2 = 1.0225 and by the formula for the period given at the end of the last subsection, and the fact that the year currently has 365.24 days, we confirm that the year has slightly less than 378 days. Square root (1.0225^3) x 365.24 = 377.64 days. This is even easier than "dot-gap counting" and much more accurate.

The reason that the earth is farther from the sun for more than half the year is that it is moving more slowly when farther away. This effect also can be seen in the last earth graph without counting dots. Note the space between the last two points with distance from the sun exactly equal to unity is approximately 4.5 of the 50 day divisions (4.60 divisions to be accurate) of the horizontal axis, but the preceding interval on the unity value line is only about three of the 50 day divisions (2.96 divisions to be accurate). This is an easier way to see that the year is essentially 378 days long with 230 days of less solar energy and 148 days with more than the average solar energy the earth had before the dark visitor changed the orbit.

The fact that the last two peaks are not exactly equal (look closely) is not a small numerical error. This is confirmed by the graphs already presented (*in Chapter 9*) in the section on accuracy. As it leaves, the dark visitor very slightly undoes some of the damage it did while approaching. While it was approaching the ecliptic from the north side, it very slightly tilted the ecliptic towards it. When it retreated on the southern side, it undid almost all of this tilting, leaving the ecliptic still tilted at approximately 23.5 degrees to the celestial equator.

Because we chose to make the intersection line (See last chapter.) identical with the X-axis, the dark visitor passes through the earth's orbital plane with Y coordinate equal to 0. If the starting Y velocity component (Vy = -0.0039 AU / hour) of the dark visitor and its starting Y distance from the sun (Y = 204 AU) are used to compute how many hours after the start the DV will "pierce" the ecliptic the result is 52,307.7 hours or 2,179.5 days of travel but the lower left corner of the second earth graph shows that the DV is closest to the earth on approximately day 2245 after the start. Initially this seemed strange so the numerical values were consulted and are reproduce in the table below (with truncation to only two decimal places make the data fit on this page).

Earth			Dark Visitor			Earth to DV	Start of
x	y	z	X	Y	Z	Separation	Day No.
-0,89	-0,02	0,37	13,40	0,46	0	14,31	2176
-0,86	-0,03	0,43	13,24	0,08	0	14,12	2180
-0,83	-0,03	0,49	13,09	-0,28	0	13,93	2184
(Sixty days omitted at this point to keep this table short.)							
0,08	-0,07	0,94	10,77	-5,91	0	12,20	2244
0,16	-0,07	0,93	10,61	-6,28	0	12,19	2248
0,23	-0,07	0,91	10,46	-6,66	0	12,20	2252

(I apologize for the fact that a comma is used in this table and in the graphs as the decimal point, but that is the style my computer uses.)

The reason why the closest approach to the earth occurs on day 2248, or 68 days after the DV has pierced the ecliptic (on approximately day 2180 where the DV's Y value changes sign) is clear with a little thought. The tilt of the XY plane with respect the ecliptic is such that on the Southern Hemisphere side of the ecliptic (DV's Y negative) the DV is moving towards the same region of space that the earth is moving to when the x and z coordinates of the earth are increasing. (x is increasing by becoming less negative) When the DV pierces, circa day 2180, z is increasing rapidly and x is becoming less negative (increasing also). Around circa 2244, x is rapidly increasing and z is only slowly starting to decrease. Thus for approximately 68 days, the tilt of the XY plane's reduction in their separation combined with the earth's motion towards the DV is more important than the fact that the DV is moving away from the earth's orbit plane, returning to deep space in the Southern Hemisphere.

I suspect that a very industrious astronomer can take all of the information I have given and obtain a significantly more accurate orientation for the XY plane and of the DV's location as a function of time. I have done this intentionally because I doubt that senior professionals in control of large telescopes will have the time to struggle through this effort. Even if they assign the task to a graduate student, who does it well, the results will still be very inferior to my knowledge of the DV's trajectory. I intend to have a fighting chance to be the first to see the DV, if it can be seen, despite the fact that I am at considerable disadvantage in trying to see it. (It will be near my horizon until it begins to seriously disturb the earth because the DV is approaching from such a high northern declination.) However, someone may do this analysis and get lucky and see it (or some shifting "twinkling stars") before me. I want whoever sees it first to have at least expended some effort, as I have. Thus I am not simply giving away my best-fit data to a more favorably located astronomer in charge of a telescope much better than mine.

If you do succeed and find the DV before I do, at least give me some credit. In some ways I want several telescopes looking for it now because the sooner mankind knows exactly what is going to happen the better we can try to prepare. Thus I do want someone to see it soon (if it can be seen optically), even if it is not me, to get highly accurate data to replace my perturbation data. To a large extent this is why I have written this report and arranged for it to appear widely rather than just send my best-fit data to the CfA/ MPC. I am currently using a pseudonym (Jack) for the reasons stated earlier (*Chapter 9, near end*) and because I fear some hostility towards the bearer of bad news. I do not want to be compelled to divulge my best-fit data by legal action (or less civil actions by the CIA etc.). I will come forward and identify myself when this report is confirmed and the initial shock subsides.

With mass 2.2 times that of the sun there is nothing man can do to deflect it[18], but with accurate data we can begin to plan a new calendar and to address the many legal problems that will be caused more calmly. For example, does your house insurance cover "loss-of-use" when the house is buried under ice? If not, how will you get enough money to move to the Southern Hemisphere? At best, there are only a couple of decades available to relocate half the world's population. At worst, the gravity gradient may break Saturn apart if the DV passes too near to Saturn and a piece of Saturn may extinguish life on earth, but to end on a more optimistic note: This event, although possible, is still extremely unlikely according to my best current data. The uncertainty as to how close the DV will come to Saturn stems mainly for the fact that the speed of the DV is still not precisely known.

---

[18] It would be easier to move the sun out of the way and everyone will agree that that is impossible.

CHAPTER 12

GRAPHICAL RESULTS

All distances in the following graphs are in Astronomical Units, AUs.

THE NEW EARTH ORBIT
378 DAY YEAR

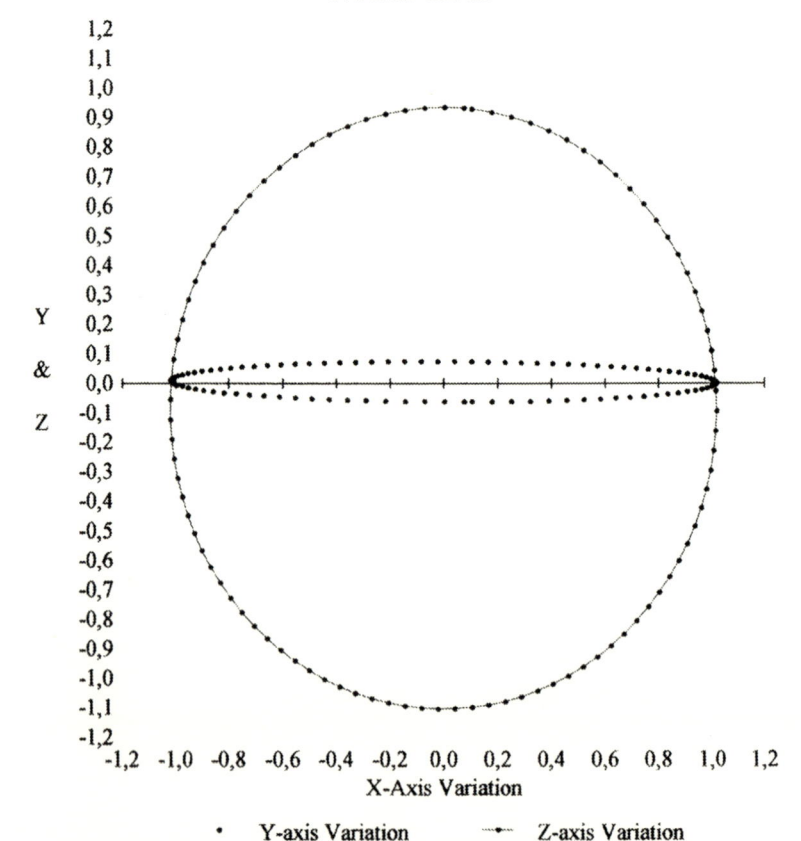

- Y-axis Variation
- Z-axis Variation

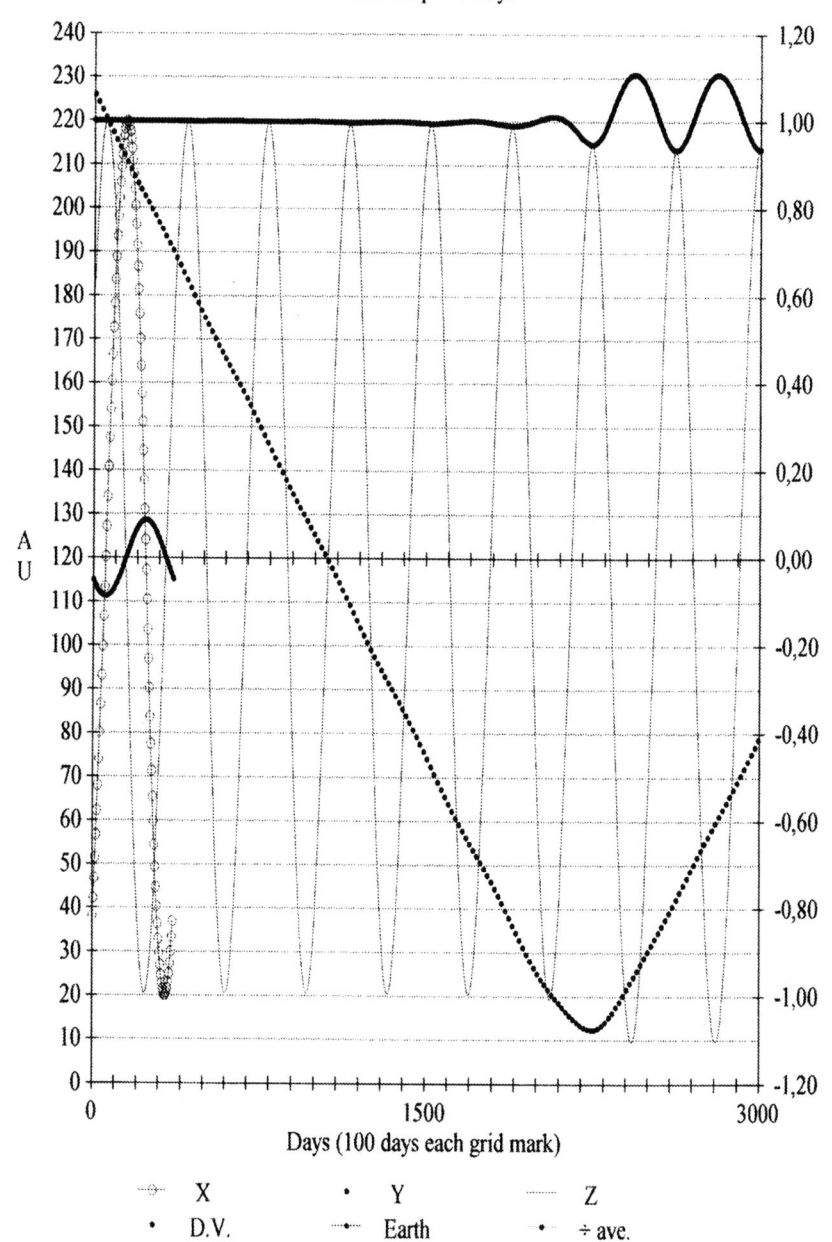

EARTH
Time Step = 4 Days

Days (100 days each grid mark)

- X
- D.V.
- Y
- Earth
- Z
- ÷ ave.

EARTH
Time Step = 4 Days

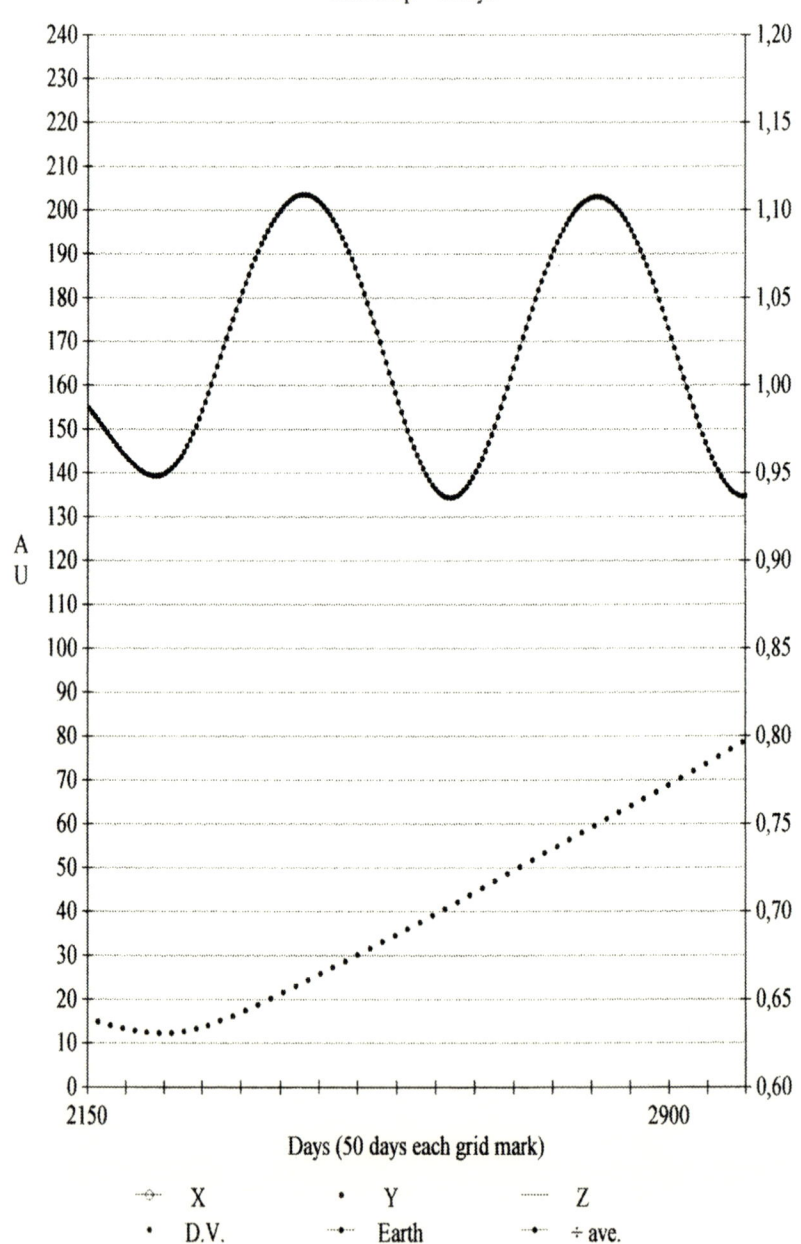

Days (50 days each grid mark)

	X		Y		Z
	D.V.		Earth		÷ ave.

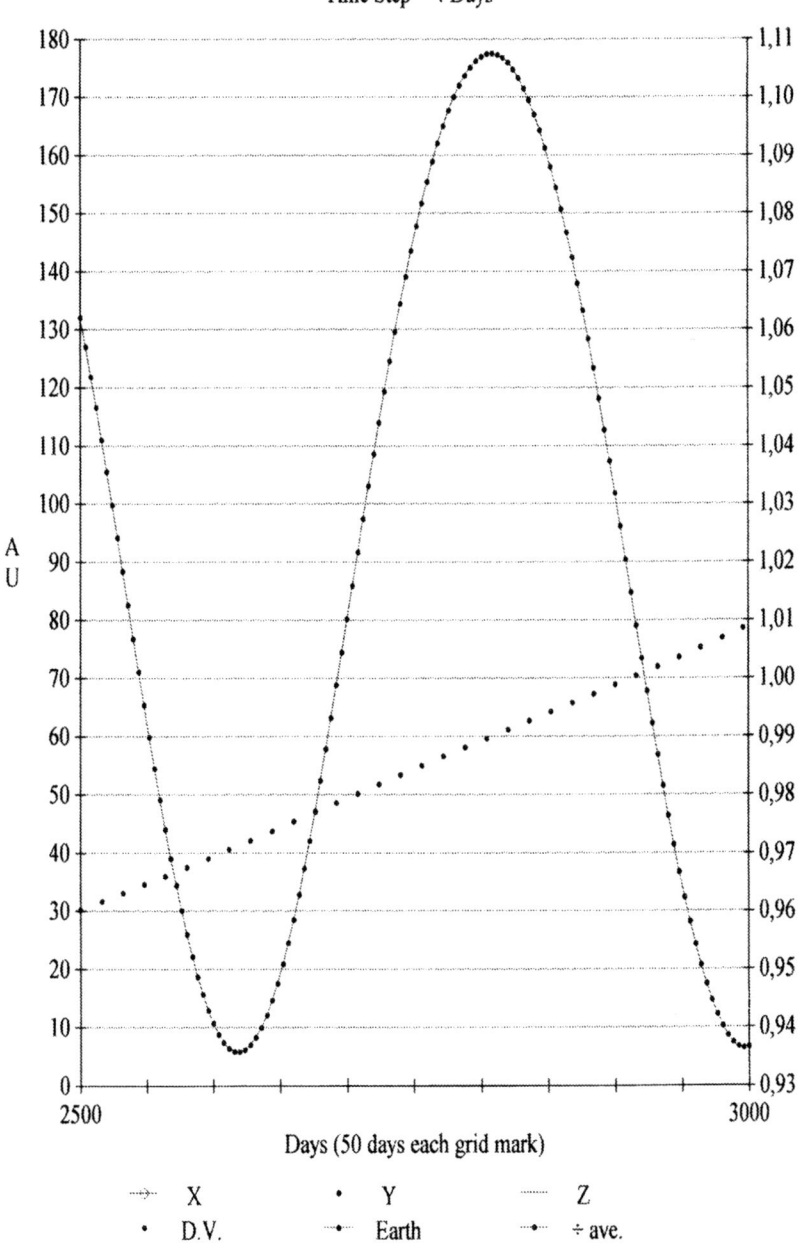

EARTH
Time Step = 4 Days

PLUTO
Time Step = 4 Days

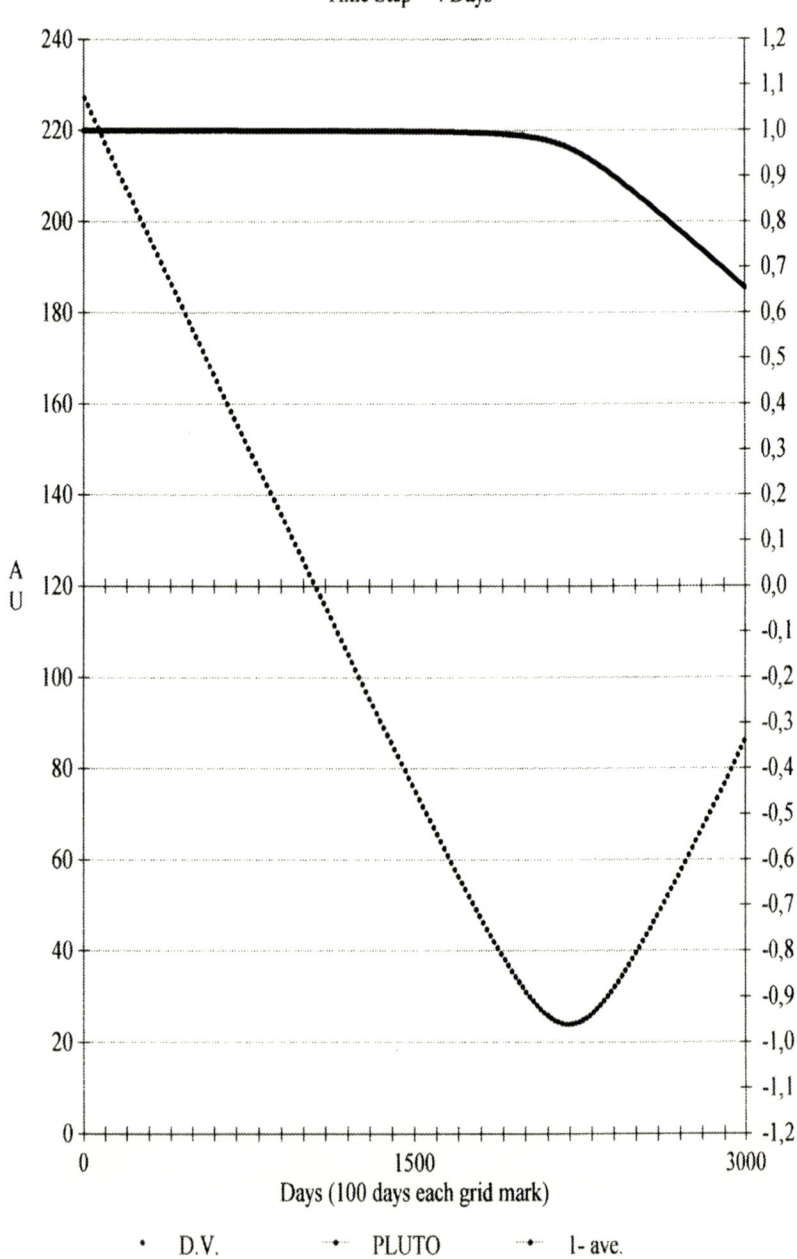

D.V. • PLUTO • 1- ave.

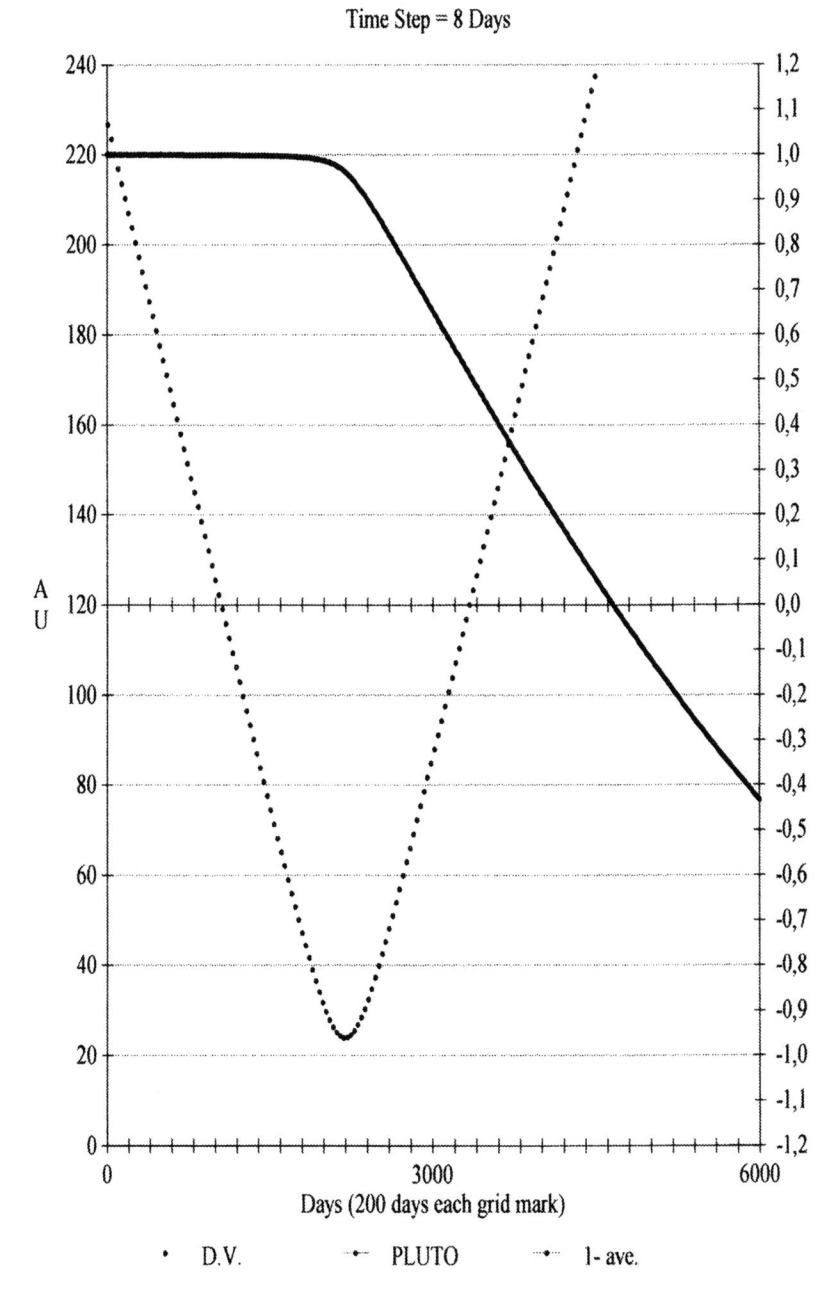

PLUTO
Time Step = 8 Days

- D.V.
- PLUTO
- 1- ave.

Days (200 days each grid mark)

Appendix Introduction

Jack finally allowed me to copy material from his hard drive for preparation of this book. I have selected only three items from it to include in this book. (I am honoring his request not to divulge anything related to the investment syndicate.) I have modified them slightly. The first subdivision of this Appendix, A1, is essentially as I found it, but I have inserted one note in it. It is an unusual record of his work when he was first trying to understand what impact a massive visitor from space would have on Pluto and latter upon the earth, when various passing trajectories were assumed, but his comments to himself recorded in his spreadsheet tell this story better than I can.

The second subdivision of the Appendix, A2, is my attempt to reorganize most of his three-body interaction model from spreadsheet rows (much too wide to print on these pages even if printed sideways) into a vertical column. I have added some notes that will help the reader understand these calculations if he wishes to build his own three-body spreadsheet model.

Both A1 and A2 were originally entries in cells of a spreadsheet. When transferred to this unstructured format, there are some awkward gaps as a result. I have corrected these whenever I could do so, but never if it changed the logical sequence of Jack's original work. Thus occasionally an equation or a defined symbol will appear out of place. (Some poorly identified, named variables, appear inside sentences).

Appendix A3 does not directly concern the Dark Visitor. After its initial boring, but essential, first paragraph, A3 compactly presents an interesting idea about black holes. Jack gives a simple proof that they contain infinite energy in zero volume, much like the big bang that started our universe. He then suggests that each black hole spawns a new universe. If this is true, the approaching Dark Visitor could be an universe!

Appendix A1

PROBLEM

Searching for the dark visitor's orbital parameters. It has already moved Pluto (and Neptune?)

If it is now "falling" towards the sun, will it soon move the earth? How long do we have?

Units, notation and physics: 3,141592654 = ¶

The visitor must be "dark" because no one has seen it, not even a Japanese "comet hunter." A "black hole" does not reflect sunlight, so the visitor could be one. Alternatively, it may be an aggregate of the "dark matter" which constitutes the bulk of the universe. Perhaps the cold dead core of a neutron star? Or something we have yet to learn of?

Analysis is in Center of Mass frame. Time is in hours. Length is in AUs. (Sun/earth distance =1.)

The XY plane is that of the sun / visitor scattering. (Planets effects on trajectories are ignored.) Planet has mass m and is at (x,y,z). Its velocity is (u,v,w). Its period is P. Its distance from Sun is R.

Visitor is at (Xv,Yv,0) with velocity (Uv,Vv,0). Visitor's mass is M times that of the sun. The visitor's distance from planet is Dvp and, and by above,

$$Dvp**2 = (Xv - x)**2 + (Yv - y)**2 + z**2.$$

It is at (Xvo, Yvo) & Sun at (-M*Xvo, -M*Yvo) initially. (The "o" designates starting conditions.)

Sun has mass Ms and is at (Xs,Ys,0). Sun's velocity is (Us,Vs,0) and its initial speed is M*Svo, where Svo is the visitor's speed at the start of detailed analysis. For consistent selection of Svo, Svo is always computed from the visitor's "cosmic speed," Svc, the speed it had when far from the sun. (Detailed analysis with different initial visitor locations have same Svc, but will have different Svo. I want to be able to compare them consistently.)

Sun's gravitational force on planet is:

$$Fs = mMsG/R^{**}2, \text{ but } R^{**}2 = (Xs - x)^{**}2 + (Ys - y)^{**}2 + z^{**}2.$$

For circular orbit: $Fs = mVt^{**}2/R = 4mR(¶/P)^{**}2$ because, $Vt = 2*¶*R/P$ where Vt is the orbit speed. For the earth: $R = Re = 1$ and period $P = Pe = 8766.24$ hours. Thus,

Vte =,000716748 AU/hour

Furthermore, in these units, by the above, $MsG = (2*¶/P)^{**}2 = 5,13728E-07$

Note MvG is just M times larger than MsG for all visitor interactions with planets.

==

Note added by Billy T: Jack is inconsistent in his notation. When he writes, he uses the decimal point in the conventional way. He lives in South America and his computer uses a comma for the decimal point. Also, for readers unfamiliar with scientific notation, the 5,137E-07 a few lines above is 0.000,000,513,7 in standard English notation. Jack allowed me to copy this program and many other things from his hard drive, but as I am a historian, I recognized that this is a document of some historical significance. It marks the beginning of a new era in man's history (assuming we survive). I have only lightly edited it, added this note and omitted many pages of the final results table. That table's right part must be folded below the left many times to fit on the page. I present only two time steps of the results table. Jack used this first model mainly to explore planet visitor / interaction as the visitor passed near the planet. Shortly before he ceased to use this model, he inserted two notes (now in italics) but did not alter or remove his original notes. All original notes were written while he was developing this first model. Unfortunately, he was less willing then to allow others to see his efforts so he inserted entirely false values for the visitor's parameters. The values given on the graphs produced with the later three body model are "essentially correct" but he is still keeping his best values secret as he want to be the first to see the visitor, if it can be seen optically.

The next appendix presents the new three-body model. Jack was so anxious to get a rough idea about the magnitude and direction of the effect of the visitor upon Pluto and the earth that he circularized all planet orbits in this model. He did however use his first approximation approach to evaluation of the average velocity and forces acting during the time step. In the results table at end, the first approximation results are in italics and below the starting condition of each time step. In his later three body model he continued to use italics for the first approximation results but put the entire time step on a single, very wide, row of the spreadsheet.

Appendix A1 makes reference an "auxiliary program," used to make sure one parameter (Svc) is consistently chosen with two other parameters (A and B). I have carefully searched Jack's computer but it is not there. He must have discarded it when he stopped using the original approach and constructed the "three body model."

End of book author's inserted note.

==

Small time-step analysis method:

If this is a scattering event (not mutually bound orbits) then both the sun and the dark visitor travel on parabolic paths in the Center of Mass coordinate system. Thus we assume a parabola for the visitor. The sun's corresponding X and Y coordinates are simply (-M) times larger. With speeds and positions known for both visitor and sun, we can calculate the force (acceleration) acting on the planet and apply it for a small increment of time to compute the velocity of the planet at the end of the time step. (Initially the planet has the same velocity as the sun plus its orbital velocity about the sun.)

The average velocity of the planet at the start and end of the time step is used to move the planet to a good first approximation of its position at the end of the time step.

The forces acting on the planet at this position are then computed and averaged with the forces already calculated for the start position to obtain a better evaluation of the accelerations acting on the planet DURING the time step. This better approximation is then used to compute the second approximation of the planet's position at the end of the time step that then becomes the start position for the next time step.

All velocities are also adjusted for the start of the next time step. The planet's new velocity was part the calculation for its new position. The velocity of both sun and visitor are increased by the same factor each time step. To prevent accumulation of small numerical errors, the current visitor speed is always computed from the cosmic speed, rather than the speed of the prior time step. The speed ratio Sv/Svc is the square root of the ratio of distances from visitor's parabolic focus. (Book I, Proposition 16, Corollary VI of Newton's "*Principles.*")

This focus is at (0, f) and it is convenient to define the cosmic speed as the speed of the visitor when passing through the point equally distant from the X axis. I.e. Svc is defined at:

(X1, f) because here the distance to the focus is simply X1 and because this point is far from the sun. I.e. no significant acceleration by solar gravity has occurred. Expressions for X1 and f will be presented in the "Analysis" section.

Next paragraph is not yet implemented. Speed changes are small for time steps < or = 24 hours. I.e. First approximation is good enough for visitor's speed change.

The velocity and position of the visitor at the start of each time step and the time-step duration are used to calculate the first approximation of the end of time-step positions for both sun and visitor. Newton's "Corollary VI" is then used again at this position to calculate the end position velocities. These start and end velocities are then averaged to compute the effective visitor velocity DURING the time step, which is then used to determine the second approximation of their positions.

This process could be repeated to yield third approximations etc., but it is better to use smaller steps to improve accuracy once the effective velocities acting DURING a time step are known.

The actual implementation of this procedure, is slightly modified by a "fudge factor" ff. The first approximation for the planet's position moves it on the tangent line away from the sun where the gravitational field is weaker, but its direction to the sun is correct. There solar gravity is calculated with the square of the separation from the sun equal to (R**2 + [Vt*Tstep]**2) instead of R**2. Thus I multiply the solar gravity calculated at the "end position" by:

ff = 1+ ([Vt*Tstep]**2 / R**2).

We can not simply ignore the computed separation from the sun and always calculate the solar gravity at a point R distant from the sun because the visitor may be moving the planet away from the sun.

The factor ff improves the approximation by correcting the separation artificially increased when we move the planet along the tangent line but does not cancel real separations that may be occur because of the visitor's gravitational attraction. This "fudge factor" may not be the optimum one.

In order to determine the correct value for ff, a "switch" Sw (with value either 0 or 1) is included in the calculation of the visitor's forces on the planet. The fudge factor is adjusted to achieve a planet "orbit" that remains with constant separation from the sun with Sw = 0. No similar factor is needed for the motion of the sun and visitor because they are constrained to move on their parabolas.

The model described above contains a small subtle error, which unfortunately accumulates with each time step. The sun was placed on a parabolic trajectory and correctly given an increment of velocity, computed by Newton's "Corollary VI," each time step. It is not the sun, but the center of mass of the sun-planet system that should be placed on the parabolic path and given this increment of velocity. If we had done this, the planet would also receive the same increment of velocity as the sun. Thus we must give the missing velocity increment to the planet each time step.—If we do not do this, the sun will slowly run away from the planet.

Prior to correcting this oversight this separation was clearly exhibited numerically when I set Sw = 0.

The reason for this separation can also be easily understood when the visitor is passing by the sun-planet system with the same X axis coordinate as the planet. At this point there is no component of the visitor's force on the planet in the X direction and the sun's gravity is just sufficient to keep the planet in circular orbit. (No acceleration for the planet along the X axis.) However, during this time step, as in all others, the sun is given the velocity increment computed with "Corolla VI."

ANALYSIS:

Both sun and visitor move on assumed trajectories, in finite steps. The visitor's assumed path is:

Yv / A = (Xv/B)**2 - 1 where A and B are selected to reproduce the observed deflections of Pluto (when this method of analysis has Pluto, not the earth, as the "planet"). Once A and B are known, we can apply this analysis to predict the effects upon the earth, however, M and S must also be known. This is difficult with only observational data from Pluto because a more massive visitor moving rapidly can have essentially the same effect as a lighter visitor moving by more slowly. The visitor's speed will be accurately known only when observational data is available from a second planet. The current inability to separately determine values for M and S does not prevent accurate prediction of the effect upon the earth. What is uncertain is how long we have. A fast, massive visitor can seriously displace the earth in less than one year! A small slow one will take longer. <u>A slow massive one permanently separates us from the sun!</u> If it is very slow and far away, I may not be alive when it gets here.

The sun is always M times farther from (0,0). Thus, the sun trajectory is:

$$Y_s = M*A - M*A*(X_v/B)**2.$$

The sun is also always traveling M times faster than the visitor. Hence the ratio of its speed at the end of a time step to that at the beginning is the same as that of the visitor's.

Value for f:

The focus of the visitor's parabola is at (0, f) where the focus is:

$$f = (B**2 - 4*A**2) / 4*A.$$

This is derived from the classical definition of a parabola (Locus of points with equal distance from a "focus" point and the "directrix" line.) This equality is easily evaluated at two particular points of the parabola. For the parabola chosen, the directrix is parallel to the X axis and below it a distance now called d.

At the point (0, -A), f + A = d - A .

At the point (B, 0), Square root of (B**2 + f**2) = d, but by the first, d = f + 2A.

Hence: B**2 + f**2 = d**2 = (f + 2A)**2 which is easily solved to give the results of the first line.

Value for X1:

The point where cosmic speed is defined is (X1,f) where X1= B**2 / 2A. (Derived by setting Yv = f) Because the sun is then at (-M*X1, -M*f) the separation is:

$$(M + 1)*sqrt\{X1**2 + f**2\} \text{ AUs.}$$

Thus, as assumed, the sun and visitor are very far apart when cosmic speed is defined.
We also assume that M*Svo << 1 and ignore relativistic effects.
(The speed of light is 7.14 AU/hr.)

VISITOR   PARAMETERS (All defined above.)

Set the 6 bold type inputs in line below. (Yvo, and f of second line are calculated, not set.)

A = **0,3**   B = **24,3**   Xvo = **0,258**   M = **1**   Svc = **0,0019**   Sw = **1**
Yvo = -0,299966182   f = 491,775

Do not set Svc inconsistent with A and B! I.e. Run the auxiliary program with chosen A and B and adjust Svc to get overlap of graph of the analytic parabola with that of the trajectory computed using Sun / Visitor gravitational equations.

At the start of detailed analysis, both visitor and sun's speeds have increase from the cosmic speed by the factor = sqrt {(B**2/2A) / sqrt(Xvo**2 + (f - Yvo)**2)} which reduces to:
sqrt {2 / (1 + [2A*Xvo/(B*B)]**2)} = 1,414213514 which is approximately = sqrt(2) because 2A*Xvo/(B*B)<<1 and its square is much smaller.

Thus, the visitor's speed at start of the detailed analysis, Svo = 0,002687006 AUs / hour and then the sun's speed, Sso = 0,002687006 This sun speed is only 0,04% of light's speed so neglect of relativity is OK.

Notes on preliminary choice of visitor parameters:

Because visitor has definitely made a very small displacement of Pluto, during recent months, and probably has had at least a cumulative effect on Neptune (comparing with the old cfa data.) I believe that the visitor is approaching the sun from Pluto's quadrant. Pluto is currently 30 AU from the sun.

Lets try Xvo =60 and B equal to 100 to begin trial and error search for its orbital parameters. Its speed of approach is still a wild guess, but no longer a WAG. Speed will become more clear if Neptune (or later Uranus) are displaced. It still too soon to be sure it is disturbing Neptune also, but it seems to be! The speed assumed above is small compared to that of light but comparable to the earth's orbital speed. This is a reasonable first guess, and can be corrected when good data on a second planet is available.

I assume the visitor and sun are now "falling" towards each other, but will miss by at least 20 AU. Thus I set A=4 and M=4. (Neither of these parameters significantly changes the time required for the visitor to approach the sun (and earth). If it is a black hole, M must be at least 2, assuming all the big bang ones have evaporated or are too small to do much, but it can't be a big one as it would have already been noticed. (I am betting it is a stellar black hole.)

Pluto is very difficult for amateurs to observe and none can accurately measure the change in position that has occurred thus far. No institution is now allocating telescope time for planetary research as NASA does it better. But if they get funding they want for a mission to Pluto, this could change—no time to waste—Got to get this done quick and publish but get it right first!

The big telescopes are all concerned with deep space. Only wide-view cameras used in asteroid searches have any chance of photographing Pluto and they are looking for new objects, not carefully measuring the locations of well known wanders. Thus, I am probably the only person in the world who now knows the visitor is coming.—If it comes within 10 AU of the earth, everyone will know!

NOTE ADDED LATER—(This analysis is now trying to predict effect upon earth) The visitor's speed is still not accurately separated from the Mass effects so I still don't know exactly where in its orbit the earth will be when it is close. Must still keep initial orbit location of earth flexible.

PLANET PARAMETERS (Three angles, defined below, & Ro)

Introduction:

This section is structured so it can be applied to any planet. I will first use it to try to find visitor's orbital parameters using my accumulating observational data on Pluto (and Neptune?). If later another planet is also disturbed, I will accurately know the orbital parameters of the visitor. Then I can apply this methodology to predict the effect upon the earth.

Definitions:

Angle OP is angle between planet's Orbit Plane and the X Y plane. It is ±90 degrees or less. (Positive OP corresponds to the visitor approaching the sun from the southern celestial sphere.)

Angle IL is between Intersection Line of these two planes and the +X axis. It is also ±90 or less. (IL is an angle in the first or fourth quadrant of the X Y plane and positive in the first quadrant.)

Note that no intersection line exists if OP=0 exactly, but to preserve these definitions, we can assume an extremely small OP in this case. (The next definition uses this line.)

Angle AP gives planet's Angular Position in the orbit plane. It has range from 0 to 360 degrees. (AP is measured from that portion of the intersection line with more negative values of Y. I also assume that the planet's normal orbital motion causes angle AP to increase with time.)

Illustrations:

If AP= 0 or 180, then the planet is in the X Y plane. If OP is not 0, z is positive in this range.

If initially OP = 0, then orbit is in the X,Y plane and remains in it as all forces are in X Y plane.

However, if OP is not 0 initially, OP can decrease towards 0 as the visitor attracts it.

Defining Initial LOCATION of the planet:

(First defined relative to the sun and later converted to C of M coordinates.)

SET (in degrees) initial values for the "planet angles" OPo, ILo, and APo (defined above) Values, in radians, appear below each as required by trig functions of this spreadsheet.

45 = OPo   -60 = ILo   30 = APo
0,785398163 radians -1,047197551 radians 0,523598776 radians

NOW SET: Planet's initial orbit radius = 1 = Ro
HENCE the planets initial coordinates, relative to the sun (xos, yos, zos) are:

0,73919892 = xos   -0,573223305 = yos   0,353553391 = zos

However, we are working in the center of mass coordinate system. Thus we must add the position of the sun. Initially the sun is at (-M*Xvo, -M*Yvo). Consequently, the planet's INITIAL COORDINATES in the C of M coordinate system are:

0,48119892 = xo   -0,27326 = yo   0,35355 = zo (All in AUs.)

Defining Initial VELOCITY of the planet:

(Again defined first relative to the sun and later converted to C of M coordinates)

Because the cube of the radius (Ro set above) is proportional to the square of the period and the planet's original period, Po, is 8766.24 * sqrt(Ro*Ro*Ro) = 8766,2400 hours; The planet's initial speed, Spo, is: 0,000716748 AU per hour.

Hence the planet's initial velocity components relative to the sun (uso, vso, wso,) are:

uso = 0,000200926   vso = 0,000529819   wso = 0,000438917

NOTE ADDED LATER—*Now, with visitor crudely known, I am trying to see how earth will be effected—Lets hope the timing is such that we are on the far side of our orbit when visitor passes by the sun.*

The sun's speed is M times that of the visitor, but we need its X and Y velocity components. The "cosmic speed", Ssc = M * Svc = 0,001900 AU / hour = the sun's speed passing through the point (-*X1, -M*f). Both the visitor and the sun have higher speeds when they are close. Detailed analysis, with small steps, starts with the visitor at (Xvo, Yvo) Here both speeds have increased by the ratio sqrt {2 / (1 + [2A*Xvo/(B*B)]**2)}.

Initially, the sun is on a parabola where slope is 2*A*(Xvo)/(B**2) = 0,000262 and the V component of sun's speed is the slope times the U component, but Ss**2 = U**2 + V**2 or Ss**2 = U**2 + (slope*U)**2 = U**2(1 + slope squared).

Hence Uso = sqrt{(Sso**2)/(1 + slope**2)} and consequently, the planet's INITIAL VELOCITY in the C of M coordinate system (uo, vo, wo) is:

uo = 0,002887932   vo = 0,000530576   wo = 0,000438917 (AUs /hour.)

Evaluating forces on the planet:

The forces acting on the planet attract it towards their sources (sun or visitor). The X, Y and Z components of these forces (Fx, Fy, Fz) bear the same relationship to the force magnitude, F, as the components of the line joining the planet to the source of force does to the length of the line segment between them, Lvp and Lsp (for visitor and sun respectively).

For example, the X component of the force from the visitor, Fxv, causes an acceleration Axv = (F / m) * {(Xv - x) / Lvp } = {(Xv - x)*M*MsG/(Lvp**3)}. It is in the +X direction if Xv > x. (Lvp was given earlier, but called Dvp.)

Note M*MsG/(Lvp**3) is a "common factor" for all three force components, which is computed and separately tabulated on the right side of the results table (along with R and other terms of interest).

Now set the adjustable correction factor, C, if automatic fudge factor, ff, is not adequate. C = 0,996 Note K, the factor in the solar force equation's "common factors," is C * ff.

Can improve K by C*(1 + (v **2 / r **2)*Tstep^2) but wait till a symmetric passing run is done.

TABLE of ACCELERATIONS, RESULTING VELOCITY & POSITIONS

(Bold type is input values, or table column names, or time step "start conditions.")
(Italic is a results of [or using] 1st approx. Normal type is note or derived value.)
Note: "Go To" cells named "Pin" and "Vin" to change inputs or return here with "top".

Set here **only** time step = **3,00**
Hence K = 0,9960046 and recall: **OPo =45   ILo = -60   APo = 30**
And parabola's parameters are: **A = 0,3   B = 24,3   Xvo = 0,258**
The DV has M=1 and Svc = 0,0019

===

Final note added by Billy T's Agent: At his point. The CD has an extremely wide, confusing folded table. Not only with the results indicated in the above table title, but also with many intermediate calculations, such as the "common factors" Jack talked about and many other values he appears to have used to check results or run special tests etc. Even after checking with a physicist friend, who knows much more than I do about these things, it is not clear what most of the entries in this table mean. To include this table would require at least 14 fold out pages. The publisher I am working with tells me this is cost prohibitive, if not impossible, so this table, which no longer contains any real results, (for example the mass M of the DV has been set to 1, etc.) has been omitted. I promised Billy T that I would publish the CD without any changes, but this has proven to be impossible. Billy T's note inserted just before the graphs in Chapter 9 states: *"Appendix A1 is a historical document of considerable significance."* Thus, it is for the discussion, not these false results that he added Appendix A1. If he ever contacts me again, I trust he will understand that I honored both my promise and our contract as best as I could.—W. R. Powell

Appendix A2

Structure of the Three-Body Model

Symbol Guide:
    Left column is the name of a variable or input value required.
    The second column gives the formula for it, and / or comments.
    A comma indicates a decimal point. Input data is bold.
    *, /, ^, & E Are multiplication, division, exponentiation, & power of 10.
    A name ending in "o" indicates a value at the start of analysis.
    A variable name ending in 1 indicates a first approximation result.
    After the first time step, all equations repeat, but subscript "o" is removed.
    Note symbols and equations are presented in blocks
        First block holds two universal constants.
        Second block holds four planet input data and one derived value.
        Third block holds coordinate transform equations. Etc.

π = 3,14159265358979
MsG = 5,1377775929984E-07 (Excessive length, but no harm done.)

Opo = Planet's "orbit plane angle" An input with range -90 to +90 degrees
Ilo   = Planet's "intersection line angle" An input with range -90 to +90 degrees
Apo = Planet's "orbit position angle" An input with range 0 to +360 degrees
Ro   = Planet distance from sun initially (At start of first time step.)
Po   = 8766,24*SQRT(Ro*Ro*Ro) (circular orbit approx. Not used for Pluto.)

Xo   = Ro*(COS(APo)*COS(ILo) - SIN(APo)*SIN(ILo)*COS(OPo))
Yo   = Ro*(COS(APo)*SIN(ILo) + SIN(APo)*COS(ILo)*COS(OPo))
Zo   = Ro*SIN(APo)*SIN(OPo)

Spo = 2*¶*Ro/Po (Pluto also requires manual over ride of next three.)
Uo = Spo*(SIN(APo)*COS(ILo) + COS(APo)*SIN(ILo)*COS(OPo))
Uo = Spo*(SIN(APo)*SIN(ILo) - COS(APo)*COS(ILo)*COS(OPo))
Wo = Spo*COS(APo)*SIN(OPo)

M = Mass of Dark Visitor (As a factor relative to sun's mass)
Xvo = Initial X location of Dark Visitor
Yvo = Initial Y location of Dark Visitor
Uvo = Initial X speed of Dark Visitor
Vvo = Initial Y speed of Dark Visitor

Tstep = Set as desired, but less than 288 (196 hours recommended = 4 days)
Tstep2 = Tstep*Tstep/2 (Often used, Calculate once.)
Tstep4 = Tstep2/2 (Often used, Calculate once.)

Start of Time-Step "loop" for first approximation results

Planet location
x = xo At start, otherwise = x2. (For Pluto, over-ride xo entry here.)
y = yo At start, otherwise = y2. (For Pluto, over-ride yo entry here.)
z = zo At start, otherwise = z2. (For Pluto, over-ride zo entry here.)

Three separations:
ps1 = $x^2 + y^2 + z^2$ (square of planet - sun distance)
pv1 = $(Xvo - x)^2 + (Yvo - y)^2 + z^2$ (square of planet - DV distance)
sv1 = $Xvo^2 + Yvo^2$ (square of DV- sun distance)

Three common (often used, calculate once) factors follow:
Cps = MsG/(ps1^1,5) (But note these change each time step.)
Cpv = M*MsG/(pv1^1,5) (Consequently, they are in each row.)
Cvs = M*MsG/(vs1^1,5) (I.e. They have relative addresses.)

Five first approximation accelerations follow:
ax1 = x*cps + (Xvo-x)*cpv
ay1 = y*cps + (Yvo-y)*cpv
az1 = z*(cps + cpv)
aXv1 = Xvo*cvs (It is not necessary to explicitly calculate the acceleration
aYv1 = Yvo*cvs of the sun because it is always - M times larger.)

Five first approximation speeds follow:

u1	= uo + ('ax1'*Tstep)/2	(Quotes are required to avoid
v1	= vo + ('ay1'*Tstep)/2	confusion with cell in row 1 of
w1	= wo + ('az1'*Tstep)/2	column ax. Etc. for other names.)
Uv1	= Uvo + (aXv1*Tstep)/2	
Vv1	= Vvo + (aYv1*Tstep)/2	

Seven first approximation locations follow:

x1	= x + 'u1'*Tstep
y1	= y + 'v1'*Tstep
z1	= z + 'w1'*Tstep
Xv1	= Xvo + Uv1*Tstep
Yv1	= Yvo + Vv1*Tstep
Xs1	= -M*aXv1*Tstep2
Ys1	= -M*aYv1*Tstep2

Three distances squared (Evaluated at first approximation end positions):

ps2	= ('x1' - Xs1)^2 + ('y1' - Ys1)^2 + 'z1'^2
pv2	= (Xv1 - 'x1')^2 + (Yv1 - 'y1')^2 + 'z1'^2
sv2	= (Xv1 - Xs1)^2 + (Yv1 - Ys1)^2

Three common factors (Evaluated at first approximation end positions):

cps2	= MsG/(ps2^1,5)
cpv2	= M*MsG/(pv2^1,5)
cvs2	= MsG/(vs2^1,5)

Five "end-of-time-step" accelerations follow:

ax2	= -'x1'*cps2 + (Xv1 - 'x1')*cpv2
ay2	= -'y1'*cps2 + (Yv1 - 'y1')*cpv2
az2	= -'z1'*(cps2 + cpv2)
aXv2	= Xv1*cvs2
aYv2	= Yv1*cvs2

Displacement of sun during time step (For translation of coordinate system):

dSx	= -M*(aXv1 + aXv2)*Tstep4
dSy	= -M*(aYv1 + aYv2)*Tstep4

Five "end-of-time-step" speeds (for start of next time step) follow:
u2 = uo + Tstep*('ax1' + 'ax2')/2
v2 = vo + Tstep*('ay1' + 'ay2')/2
w2 = wo + Tstep*('az1' + 'az2')/2
Ux2 = Uvo + Tstep*(aXv1 + aXv2)/2
Vx2 = Vvo + Tstep*(aYv1 + aYv2)/2

Five "next-time-step" positions (Used as were x, y, z, Xvo & Yvo.):
x2 = x + Tstep*(uo + 'u2') / 2 - dSx
y2 = y + Tstep*(vo + 'v2') / 2 - dSy
z2 = z + Tstep*(wo + 'w2') / 2
Xv2 = Xvo + Tstep*(Uvo + Uv2) / 2 - dSx
Yv2 = 'Yvo + Tstep*(Vvo + Vv2) / 2 - dSy

(Subtracting dSx and dSy in the X and Y coordinates of the planet and DV effects a pure translation of coordinate systems in the XY plane.) The sun is simply assumed to be at the origin when the next time-step starts. (No need to explicitly calculate Xs2 - dSx = 0 etc.)

Usage (as explained to me by Jack):
All formulae above are in row one of the spreadsheet. They are copied down to the second row with minor changes. Specifically, all terms that ended with "o" are changed to refer to the corresponding terms just computed in the first row. Typically these are the terms that end in "2" just above. Once this second row is correctly written, the third row is made from it without any thought just by copying the full row down as most cell references in the formulae are relative addresses. Etc. for as many time steps as desired, up to the limit of the spreadsheet. A few terms in the calculation, like the value of ¶, M and MsG, etc. do not change. These are listed once and their absolute address is used in each row to permit the "copy down." Despite the earlier use of the word "loop" to describe this structure, there is no loop logic to construct and one does not need to worry if one made the correct number of "passes" through the loop as one did with Fortran type logical code etc. Already mentioned in the main text, is the fact that all the intermediate positions and speeds are available for graphical analysis and testing. Spreadsheets provide error free and flexible plotting subroutines. They should be used more for scientific work of modest size to avoid errors.

Appendix A3

While Jack was using my computer to make his final review of my book, I was again looking at files in his portable. I found an interesting one, related to black holes, that I had missed earlier in part because Jack had called this file "ZenoSteps." I mistakenly thought it had nothing to do with the Dark Visitor and initially only read the first paragraph, which seems to be making some boring calculation about fundamental particles. Here it is:

ZenoSteps:

A fundamental particle of mass m is, with many others, moving to a point. Consider it when it is R from the point that it and the others are moving towards. Also assume that many others are already closer to that point so that this particle currently has weight w due the attraction of the others that are closer. The work done on it by their gravitational field as it moves to position r, a small fraction f =(R-r)/R of the way closer towards the center, is greater than w(R-r) = wRf because its weight is increasing as all particles move closer to the central point. At r, its weight has increased slightly to W = w(R/r)^2. Now as it moves still closer to the center by the same fractional amount or thur a distance d = rf, the work done on it is greater than Wd = Wrf = {w(R/r)^2}rf = wRf(R/r) > wRf. That is, the work done on it during the second fractional increment of its move towards the center is slightly greater than that done during the first fractional move. This is because the distant moved is linearly related to its current location but the force of gravity is increasing quadraticly as it, and the others that were already closer, approach the point.

The same argument shows that work done during the third move towards the point by the same fraction of the distance remaining is also slightly greater than that of all prior such fractional moves. The particle can continue to move fractionally closer to the point and never reach the center in a finite number of moves. Thus the work that has been done on it when it is at the point is infinite.

That is, the average energy in each particle in all black holes is infinite. They must be infinitely hot. Infinite energy concentrated in zero volume sounds very much like the "big bang" that started our universe. Perhaps every black hole is spawning a new universe.

The conclusion that black holes have infinite energy concentrated in zero space is unchanged even if space is quantized so that it is impossible to move <u>continuously</u> thru it. I.e. once the particle is some very small distance, d, from the central point of the black hole, then further fractional movement towards the center is impossible (or a meaningless concept) because the next fractional step, d, is less than d_{min}, the distance between adjacent, but discrete, points of space. I.e. there are only a finite number of steps towards the center. However, if integration, not finite steps, were used to evaluate the work done, only this last step still does infinite work. Thus quantized space only blocks Zeno's computational approach, but does not change the conclusion that black holes have infinite energy in zero volume.

One could assume that gravity is not an inverse square law force on a very small scale, so that the final movement to the central point in this anomalous gravity region adds only a finite amount of energy to each particle, but they should still be very hot if all collapse to one point, perhaps hot enough to still spawn a universe. With this modified gravity assumption, our universe does not have infinite energy in it, only an inconceivably large, but finite amount. The many universes it spawns via black holes must each have much less than we do.—Except for the very short intervals permitted by the uncertainty principle, I believe that energy is conserved. Thus, each universe in the chain must be much more limited than the prior ones. I wonder where we are on the chain?

Postscript

Anonymous but creative writers of computer programs like to hide "Easter Eggs" in their work. We (Billy T, and Jack) have done the same. One, <u>very</u> near the front of the book, is compact. It becomes evident when you change your perspective. Another is a very small error in one or more of the graphs. It is just detectable if you change your perspective again when looking at graphs. Still a third is diffuse and subtler. It was created when Jack changed his perspective. It is called diffuse because some calculations made and illustrated accurately in tables etc. directly contradict statements of a truth made several times elsewhere in the book, but neither the calculations nor these statements are the egg's location. This egg is located in an Appendix where it isn't, but should be. This third described egg is a missing item, an analysis design flaw, like those intentionally placed in high-quality Ukrainian Easter Eggs as a hidden signature or to not offend God by pretending to perfection. The existence of this flaw causes the conflict between text and calculations. X marks the spot of another.

There are others, about which we are not giving hints or describing. One is <u>very</u> near the main division of the book. It also is an intentional error—something real and inherently dynamic is described in verbal terms that are inappropriate and physically impossible, but better understood by the average reader (one with only modest knowledge of physics) if expressed as it is. You are told something about most of the eggs so you will know when you have found them. Anyone finding five or more eggs should contact Billy T or Jack when they "come forward" to publicly identify themselves if the egg hunter wish to receive a signed certificate, suitable for framing, of this fact. It is the "World-Class Egg Hunter" certificate. (This is a solitary game you play and you will learn some interesting facts if you play it. A stamped self-addressed envelope and one dollar handling fee will be required.) The first egg is relatively easy to find, only a reading / perception trick, but you must think to find the others. In general, the closer to the end of the book you come the harder the eggs are to find, but even readers without a scientific background can painlessly (and we hope entertainingly) learn enough physic to find them all if they carefully read the entire book. (Don't spoil the hunt for other readers by telling where they are.) These errors are included in the book to encourage you to critically study some parts that you probably skipped during the first reading of this disguised didactic book.

0-595-30364-1

Printed in the United States
75767LV00004B/195